IRREVERSIBLE THERMODYNAMICS

IRREVERSIBLE THERMODYNAMICS
Theory and Applications

K. S. Førland, T. Førland and S. K. Ratkje

The University of Trondheim
The Norwegian Institute of Technology
Department of Chemistry

JOHN WILEY & SONS

Chichester · New York · Brisbane · Toronto · Singapore

Library of Congress Cataloging in Publication Data:

Førland, K. S. (Katrine Seip)
Irreversible thermodynamics.
1. Irreversible processes. I. Førland, T. (Tormod)
II. Ratkje, S. K. (Signe Kjelstrup) III. Title.
QC318.I7F67 1988 541.3'69 87–25408

ISBN 0 471 91706 0

British Library Cataloguing in Publication Data:

Førland, K. S.
Irreversible thermodynamics : theory and applications.
1. Irreversible processes
I. Title II. Førland, T. III. Ratkje, S. K.
536'.7 QC318.I7

ISBN 0 471 91706 0

Typeset by Photo·graphics, Honiton, Devon.
Printed in Great Britain by Biddles Ltd., Guildford.

CONTENTS

PREFACE

Irreversible thermodynamics is an extension of classical thermodynamics to give a unified method of treating transport processes. It is particularly useful when several interacting transport processes occur simultaneously. Simultaneous transport processes, e.g. transport by diffusion and electric current or the transport of heat coupled to transport of mass or electric charge, are encountered in many fields of science. These are common phenomena in different fields such as electrochemistry, chemical engineering, biochemistry and biophysics, metallurgy and geology.

The book is intended for senior level students, but may also be useful to others who intend to apply the principles of irreversible thermodynamics to their problems. To facilitate understanding, the basic equations are derived in a simple way using a minimum of mathematics. Equations of transport are therefore derived for transport in one direction only. Furthermore, the derivations are first carried out for very simple systems and then extended to more complex cases. In order to develop a physical understanding of the mathematical operations, the quantities used are operationally defined, i.e. experimental determinations of the quantities are described or outlined.

The symbols are in agreement with the IUPAC recommendations and SI units are used. We have made efforts to keep the language simple, avoiding — as far as possible — specialized terminology in the different fields.

In Part I of the book the theoretical fundamentals are developed and related to reality by examples. In Part II the theories are applied to solve some important problems within varied fields of science and technology. The chapters of Part II can be read independently of one another.

Many colleagues from several countries have been helpful to us with suggestions and advice leading to improvements in the text. They are too numerous to be mentioned by name, but we are grateful to them all for the keen interest they have shown. We are also indebted to our students who have scrutinized the manuscript for mistakes and obscurities.

Katrine Seip Førland *Tormod Førland* *Signe Kjelstrup Ratkje*

PART I

Theory

CHAPTER 1

Introduction

Classical thermodynamics deals with the driving forces for reactions and with *equilibrium*. It is well established as one of the pillars of the natural sciences. The theory of equilibrium, however, is limited to systems that can be regarded as isolated from the environment. It is not applicable where there is transport of heat, electricity or matter across the borders of the system. Open systems with transport across the borders are very familiar. The steady influx of energy from the sun to the earth counteracts the formation of equilibrium. Changes and movements are more common than situations of equilibrium where no macroscopic change takes place. *Irreversible thermodynamics* was developed to take care of *non-equilibrium* situations. Irreversible thermodynamics also has its limitations, and the treatment in this book is limited to near-equilibrium situations with *microscopic reversibility*, that is *local equilibrium*, and *linear transport processes*.

It is customary to date the beginning of irreversible thermodynamics back to Thomson's work in 1854 on the thermoelectric effect, interacting transports of heat and electric charge.[1] The consecutive studies of phenomena and development of theory were undertaken by some of the world's most famous scientists: Helmholtz, Boltzmann, Nernst, Einstein, to mention just a few. The papers by Onsager[2, 3] on *Reciprocal Relations in Irreversible Processes*, Parts I and II, in 1931 constituted an important milestone in the development of the theory. In the following decades the theory was further developed, and the assumptions behind the theory were scrutinized. Some leading names in these studies are Meixner, Prigogine, and de Groot and Mazur.

The main objective of irreversible thermodynamics is to describe completely and quantitatively interacting transport processes.

Simpler problems of interacting transports can be treated without the use of irreversible thermodynamics, but the unifying theory is an important tool for organizing more complex problems. Irreversible thermodynamics gives the set of equations needed to deal with simultaneous transport processes in a systematic way. The conditions of a system often imply relations that can reduce the number of parameters needed to describe the process. The systematics of irreversible thermodynamics helps to analyse these relations. Thus the number of independent experiments needed to study the system can be reduced. Alternatively a consistency check can be made by determining the same quantity by different methods.

The theory of irreversible thermodynamics is well suited for the study of transports in electrolyte systems (see Chapter 4). It adapts itself to electrolyte solutions and to fused salts. The theory is almost indispensable for transport processes through membranes (see Chapter 5). The use of the theory for studying biological phenomena is steadily increasing. With the increasing use of membranes in technological processes, the characterization of transport parameters and their interrelation is of interest. Transport of heat coupled to transport of matter and electric charge (Chapter 6) and the influence of gravity on transport phenomena (Chapter 7) are generally better understood by the use of irreversible thermodynamics.

To demonstrate the applicability of irreversible thermodynamics to a variety of practical problems, some examples are given in Chapters 8–11. The reader who works with transport problems may find other applications of the theory.

1.1. *The Dynamic Equilibrium*

The second law of thermodynamics gives the fundamental equilibrium criterion, the sum of all entropy changes for the system and the surroundings is equal to zero, $dS = 0$. For an irreversible process this sum has a positive value, $dS > 0$. The equilibrium state is a *dynamic state*, where the rates of processes in opposite directions are equal. When a system is slightly out of equilibrium, and it approaches equilibrium, processes in opposite directions are taking place continuously. The rate in one direction is slightly higher than the rate in the opposite direction until equilibrium is attained and the rates become equal. This description is valid both for chemical reactions without any coordinates in space and for directed transport processes.

The transport processes can be illustrated with an example. Consider the electrochemical cell:

$$(Pt)Cl_2(g)|AgCl(s)|Ag(s)$$

The charge carriers in the solid electrolyte are Ag^+ ions moving over interstitial positions. When the cell is open, or the emf of the cell is balanced by an outer emf, there is no net current. With no net current the electric potential over the electrolyte is zero. The charge carriers are still in motion, but an equal number of steps are taken in each direction, and they cancel. The movements of interstitial Ag^+ ions can be pictured as going from one interstitial position to another over an energy barrier, E_a (see Fig. 1.1(a)). No direction is preferred.

Due to vibrations an Ag^+ ion in an interstitial position will have a probability of reaching the top of the barrier proportional to an exponential function, $\exp(-E_a/kT)$, where E_a is the activation energy, k is the Boltzmann constant and T is the absolute temperature. There is no net transport in any direction. The rate of transport to the right, $v_{\rightarrow}$, is equal to the rate of transport to the left, $v_{\leftarrow}$. At equilibrium $v_{\rightarrow} = v_{\leftarrow} = v_o$, where v_o can be expressed as

$$v_o = k_1 \exp(-E_a/kT) \tag{1.1}$$

where k_1 is a constant.

Figure 1.1(b) shows the situation when the emf of the cell is not balanced, i.e. a net current passes from left to right. There is now an electric potential, $\Delta\varphi$, between adjacent interstitial positions, and the height of the barrier is reduced to $E_a - \frac{1}{2}e\Delta\varphi$ for movements to the right, while the barrier is increased to $E_a + \frac{1}{2}e\Delta\varphi$ for movements to the left, where e is the charge of an electron. The net rate of movement from left to right is

$$v = v_{\rightarrow} - v_{\leftarrow} = k_1 \exp(-E_a/kT)\,[\exp(\tfrac{1}{2}e\Delta\varphi/kT) - \exp(-\tfrac{1}{2}e\Delta\varphi/kT)] \tag{1.2}$$

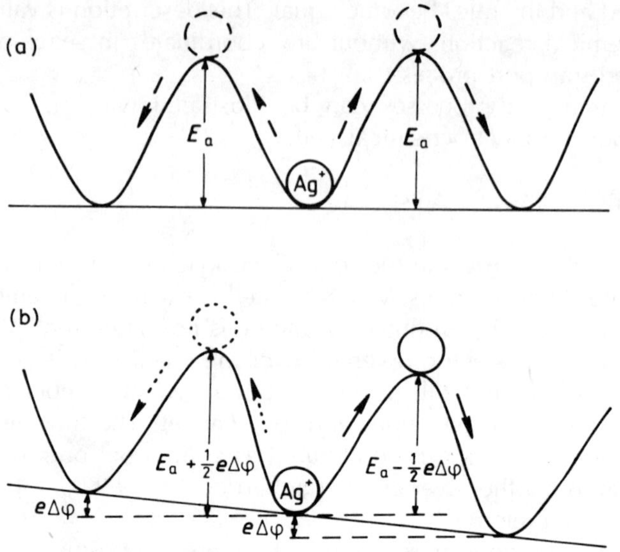

Figure 1.1. Motion of an ion from one position to a neighbouring position over an energy barrier. (a) No electric field. The energy barrier, E_a, is equal in both directions. (b) With an electric field. The energy barrier is decreased by an amount $\frac{1}{2}e\Delta\varphi$ in the direction of the field and increased by the same amount in the opposite direction.

1.2 *The Linear Range of Transport Processes*

Equation (1.2) is an exponential equation, and it does not easily lend itself to treatment by irreversible thermodynamics. For small electric potentials, however, where $\frac{1}{2}e\Delta\varphi \ll kT$, the approximation $e^x \approx 1 + x$ can be used; v_o can also be introduced from eq. (1.1), and from eq. (1.2) a simplified expression is obtained for the net rate of movement:

$$v = v_o\,(e/kT)\Delta\varphi \tag{1.3}$$

The exponential rate equation is thus reduced to a linear relation between the rate, v, and the force, $\Delta\varphi$. The force is expressed as the electrical potential over a distance Δx of molecular dimensions. This may also be expressed as the ratio $\Delta\varphi/\Delta x$ or $d\varphi/dx$. Forces can be of different kinds. An example is a change in chemical potential with distance, $d\mu_i/dx$, which leads to diffusion of matter.

Linear relations, such as eq. (1.3), are valid for processes where the net rate of change in one direction is small compared to the two rates in opposite directions (v is a small difference between two large values $v_{\rightarrow}$ and $v_{\leftarrow}$), i.e. there is a *microscopic reversibility*. It can be assumed that thermodynamic state functions have the same values as they would have in an equilibrium situation, i.e. there is *local equilibrium*.

Prigogine[4] carried out theoretical studies on the limitations for the validity of thermodynamic state functions when a system is not in equilibrium. He concluded that the fundamental Gibbs equation, eq. (2.3) (see Chapter 2), is valid for a large class of irreversible processes.

One can find examples of rate processes without a microscopic reversibility. In these cases the approximation $e^x \approx 1 + x$ cannot be used in the equation for the net rate. One example is the discharge process at the righthand-side electrode of the given electrochemical cell, (Pt)Cl_2(g)|AgCl(s)|Ag(s), for higher current densities. From the theory of *electrode overpotential*[5] an equation is obtained for the current density:

$$j = j_o\,[\exp\,((1-\beta)e\eta/kT) - \exp\,(-\beta e\eta/kT)] \qquad (1.4)$$

This equation resembles eq. (1.2). The energy barrier for the transport, however, may be asymmetric. This is expressed by β, which for a symmetric barrier is equal to $\frac{1}{2}$. The *exchange current density*, j_o, is analogous to v_o. The *overpotential*, η, is analogous to $\Delta\varphi$. An experimental value for j_o can be obtained from measurements when $e\eta \ll kT$, in the linear range of eq. (1.4), where we have

$$j = j_o\,(e/kT)\eta \qquad (1.5)$$

which is similar to eq. (1.3). For large numerical values of the overpotential, when $e|\eta| \gg kT$, one of the exponential terms in eq. (1.4) will predominate over the other, and the process will virtually occur in one direction only. Linear relations are then no longer valid. We shall not go further into the theory of overpotential.

Linear transport laws are in frequent use. Apart from the example given in eq. (1.5) there are *Fick's law of diffusion, Fourier's law of heat conduction, Ohm's law of electric conduction* and other linear relations between a force and a flux. We shall seek a common basis for these linear laws and study the interaction between different processes.

1.3 *The Concept of Components*

The choice of components to describe a system is a central point in this treatment of transport processes. The number of transport equations and the kind of forces used in the calculations depend on the choice of components. The following definition will be used:

> *'The number of thermodynamic components in a system is the minimum number of neutral chemical species necessary to describe any part of the system.'*[6]

This is often called the number of components in terms of *Gibbs Phase Rule*.

Let us consider a system with a number of species present. When some of these are related by chemical equilibrium or stoichiometric conditions, the number of species needed to describe the system is reduced by one for each relationship. The number of components is the number of species less the number of relations between the species. Any more than this minimum number of species will overdetermine the system.

In this description of the system and the changes taking place in the system, a choice may be made between several sets of components as long as they fulfill the above definition. Measurable thermodynamic properties are independent of the choice of components. The same applies to a section of the system. Thus one is permitted to use different sets of components in different sections when describing changes in thermodynamic properties.

A chemical equilibrium implies a relation between the chemical potentials of the species involved because the Gibbs energy is equal to zero:

$$\Sigma_i \nu_i \mu_i = \Delta G = 0 \qquad (1.6)$$

Here ν_i is the stoichiometric coefficient for species i (negative for reactants, positive for products). For a system of, for example, the three species H_2, O_2 and H_2O in an established equilibrium, $2\ H_2O \rightleftarrows 2\ H_2 + O_2$, the chemical potentials of the three species are related:

$$2\ \mu_{H_2} + \mu_{O_2} - 2\ \mu_{H_2O} = 0$$

Thus any two of the three species can be considered as the components of the system. Unless at very high temperature, the equilibrium is shifted far to the left. A reasonable choice of components would be H_2 and H_2O, when there is an excess of hydrogen over the stoichiometric amount to form water, and O_2 and H_2O when there is an excess of oxygen. In the case of a system with variations in composition over a range involving excess of hydrogen on the one side, and excess of oxygen on the other side, it may be inconvenient to change from one set of components to another. In such a case one may prefer to use the components H_2 and H_2O over the whole range. This will mean that mixtures high in oxygen content are obtained by adding H_2O and removing H_2. In the calculations a negative amount of H_2 is entered. The *negative content of a component* is a formal concept. It does not create any problem in thermodynamic calculations.

When a compound is stable to decomposition under all the conditions of the system, the decomposition reaction and its products are ignored. For a system containing liquid water (or water vapour at a moderate temperature), the decomposition into hydrogen and oxygen is not considered.

Ions are not components. According to the definition above only *neutral* species are considered as components. Under no circumstances is it possible to vary the amount of only one kind of ion without violating the electroneutrality principle. Even a small excess of ions of like sign would mean a substantial electrostatic energy. For ion-containing systems, salt melts or solutions of electrolytes, transport processes create only negligible changes in the electrostatic energy of the system.[7] To deal with these systems, however, it is customary to introduce chemical potentials for single ions and local electric potentials, which are undefined and unmeasurable quantities. In this book we shall avoid these quantities (see Chapter 4 for a discussion of the problem).

The treatment of transport processes in this book is limited to situations of linear relationship between fluxes (rates) and forces.

Local equilibrium is established anywhere in the system. When the transport equations are based on the minimum number of neutral components, independent forces are obtainable. As we shall see later (Chapter 5), the independence of forces is a condition for some relations between coefficients.

1.4 *Coupling of Transports in Membranes*

Transport through membranes is of central importance in biologial processes, and of increasing importance in science, technology and industry. In many cases the transports of different species are coupled, i.e. the species do not move independently. One example is water of hydration, which is transported together with the hydrated species. In some cases the coupling is complete, i.e. the one species cannot move without the simultaneous movement of the second species, and vice versa. We shall look at models for two different types of coupled transport processes.

Let us consider a membrane separating two solutions, I and II, of mixtures of the neutral solutes A and B. Solution II has the higher concentration of both solutes. Suppose that A and B are transported through the membrane only when forming a complex with a carrier molecule C in the membrane, and that C can migrate only when forming a complex with A or B. The species C cannot leave the membrane. The system is shown in Fig. 1.2.

In the stationary state any net transport of A across the membrane is compensated by a net transport of B in the opposite direction. The transfer of A from solution II to solution I is coupled to the transfer of B from solution I to solution II by the two diffusion processes of AC and BC across the membrane. A net transfer of A from high to low chemical potential for A, forces B to move against the chemical potential difference for B. The driving force

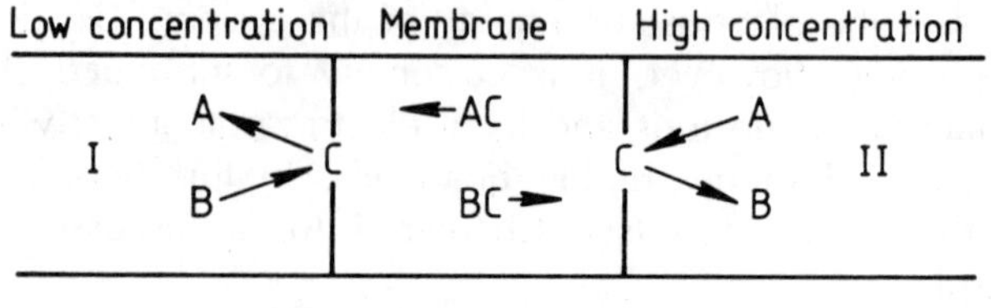

Figure 1.2. A carrier, C, transports solutes, A and B, across a membrane.

for the transport of A from right to left is thus ($\Delta\mu_A - \Delta\mu_B$). The requirement for this process to take place is that $\Delta\mu_A > \Delta\mu_B$. If $\Delta\mu_A < \Delta\mu_B$, there will be a net transport of B from high to low chemical potential, and A will be forced to move in the opposite direction.

Even with large differences in concentration between solutions II and I, the driving force for the diffusion ($\Delta\mu_A - \Delta\mu_B$) may have a small numerical value compared to RT (or kT if μ refers to chemical potential per molecule). This means that the transport may be described by linear equations.

For systems of electrolyte solutions separated by an ion exchange membrane a similar situation may exist. This time let us consider a cation exchange membrane separating two solutions, I and II, of mixtures of the salts AX and BX. As before, solution II has the higher concentration of both solutes. The salts are completely dissociated into A^+, B^+ and X^-. When the membrane is purely cation selective, the cations A^+ and B^+, but not the anion X^-, can migrate through the membrane. The system is shown in Fig. 1.3.

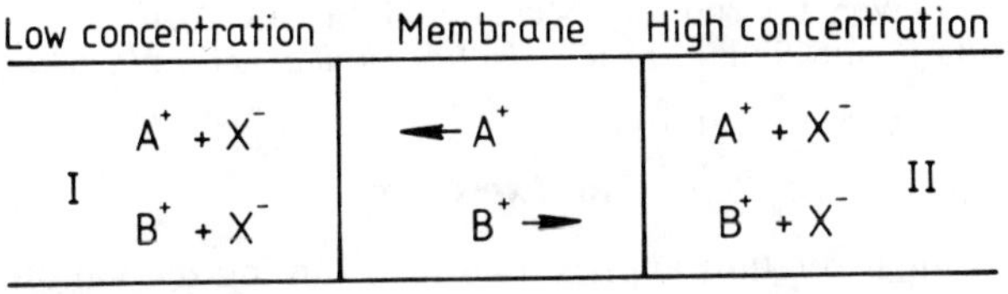

Figure 1.3. A cation exchange membrane, through which the positive ions, A^+ and B^+, can migrate.

In this case the coupling of the transfers of A^+ and B^+ is due to the requirement of electroneutrality throughout the system. A net transfer of A^+ from high to low chemical potential for AX forces B^+ to move against the chemical potential difference for BX and the driving force for the transport of A^+ from right to left is again given by a difference in chemical potential ($\Delta\mu_{AX} - \Delta\mu_{BX}$).

Biological systems frequently have large concentration differences for a component across membranes with a thickness of about 10 nm. When transports are coupled by some mechanism of the types described above, the driving force for diffusion may nevertheless be small, giving linear relationships between fluxes and forces. The whole process may take place at near-equilibrium conditions with microscopic reversibility. Thus the transformation of chemical energy into work may be very efficient.

Many processes in biology and in technology involve coupling of transport and a chemical reaction. The chemical reaction leads to transport of a component against a gradient in chemical potential. This is called *active transport*. The force of a chemical reaction, the affinity, is a scalar with no direction in space. The coupling of this scalar to a vectorial transport process is treated in Sections 8.3 and 10.1.

1.5 *References*

1. Thomson, W. (Lord Kelvin), *Proc. Roy. Soc. Edinburgh,* **3**, 225 (1854).
2. Onsager, L., *Phys. Rev.,* **37**, 405 (1931).
3. Onsager, L., *Phys. Rev.,* **38**, 2265 (1931).
4. Prigogine, I., *Physica,* **15**, 272 (1949); *J. Phys. Chem.,* **55**, 765 (1951).
5. Bockris, J. O'M. and Reddy, A. K. N., *Modern Electrochemistry,* Plenum Press, New York, 1970.
6. Lewis, G. N. and Randall, M., *Thermodynamics* (revised by Pitzer, K. S. and Brewer, L.) 2nd ed., McGraw-Hill,, New York, 1961.
7. Førland, T. and Ratkje, S. K., *Electrochimica Acta,* **26**, 649 (1981).

1.6 *Exercises*

1.1. *The linear relation approximation*. Examine the validity of the approximation, $e^x \approx 1 + x$. Calculate the percentage error when replacing e^x with $1 + x$ for values of x: (i) 0.001, (ii) 0.01, (iii) 0.1, (iv) 0.5.

1.2. *The concept of components*. (a) Given a solution of sodium chloride in water: (i) What species are present in the solution? (ii) Give the relations between amounts of the species. (iii) Which are the thermodynamic components? (iv) Give an expression for the mole fraction of each component. What is the sum of the mole fractions? (v) How many freely variable mole fractions are there? (b) Given a solution of sodium chloride and hydrogen chloride in water, answer the questions (i) to (v) above for this solution. (c) Given a solution of a weak acid, HA, in water, answer the questions (i) to (v) above for this solution.

1.3. *Coupling of transports in a membrane*. Two aqueous solutions of equal volume are separated by a purely cation-selective membrane. Initially the *lefthand-side* solution contains 0.1 kmol m^{-3} KCl *and* 0.1 kmol·m^{-3} HCl, while the *righthand-side* solution contains *only* 0.1 kmol m^{-3} KCl. (a) Give the directions of net diffuson for K^+ and H^+ in the membrane. (b) After some time, but before equilibrium has been attained, give the driving force (expressed by chemical potential differences) for the diffusion of H^+ from left to right. What is the driving force for the diffusion of K^+ from right to left? (c) Give the condition for equilibrium in terms of differences in chemical potentials. Calculate the concentrations of KCl and HCl on both sides of the membrane at equilibrium, assuming ideal solutions. For ideal solutions, $\Delta\mu_{HCl}$ = $RT\ln\,[c_{H^+,\ell}c_{Cl^-,\ell}/(c_{H^+,r}c_{Cl^-,r})]$, and similar for $\Delta\mu_{KCl}$; ℓ is left-hand-side solution and r is righthand-side solution. Neglect any transfer of H_2O.

CHAPTER 2

The Entropy Production in Irreversible Processes

The characteristic distinguishing an irreversible process from a reversible one is an increase in the total entropy, while there is no change in the total entropy for the reversible process.

This *entropy production* in an irreversible process leads to a *dissipation of energy*. The dissipated energy is converted to heat, which is not available for carrying out work. In a reversible process there would be no dissipated energy.

In an irreversible process there may be a flow of heat, mass and/or electric charge. When studying the changes taking place in irreversible processes, we need to consider these flows and their driving forces. *The forces* are given by *variations in thermodynamic functions*, e.g. temperature, chemical potential and electric potential. The flows are usually described by *fluxes* — the amounts of heat (or entropy), components and charge crossing a plane of unit area in unit time. The plane is oriented perpendicular to the fluxes and its position is given by an external coordinate system — a *frame of reference*.

Equations will be developed for the entropy production and for a *dissipation function* when local equilibrium can be assumed. These equations can be interpreted as products of forces and fluxes.

The internal energy: the change in the internal energy for a *closed system*, by any process, is defined by the change in *external variables*:

$$dU = dq + dw \tag{2.1}$$

Since heat, q, and work, w, are not state functions, many alternative paths can afford the change dU. One may choose a reversible path where $dq = TdS$ and $dw = -pdV$, and thus dU can be expressed by internal variables:

$$dU = TdS - pdV \tag{2.2}$$

In eq. (2.2) the change in internal energy is expressed by state functions only. The change in a state function is independent of the path and eq. (2.2) is *generally valid, regardless of by which path the change in internal energy was brought about*. For a closed system $U = f(S,V)$; thus dU in eq. (2.2) is a *total differential of internal variables*.

For an open system the internal energy is, in addition, a function of the amount of matter, $U = f(S,V,n_1, n_2, \ldots)$, where n_i is the amount of component i. The change in internal energy, expressed as the total differential of internal variables, will then be

$$dU = TdS - pdV + \Sigma_i \mu_i dn_i \tag{2.3}$$

where

$$\mu_i = (\partial U/\partial n_i)_{S,V,n_{j \neq i}}$$

is the chemical potential of component i. The subscript i symbolizes the number assigned to one term in a series. The summation of all terms in the series is symbolized by Σ_i. Here it means that the summation is over all components. When there is a need to distinguish between two terms or two series of terms, the symbol i will be used for the one and j for the other. Equation (2.3) contains only state functions, dU is a total differential of internal variables and the equation is generally valid for an open system. Chemical potentials do not have absolute values, and the value of dU for an open system can only be given with respect to a reference state.

In the derivations in this book electric potentials will enter the calculation via the first law as *external* electric work.

2.1 *Entropy Production in a Discontinuous Adiabatic System*

Let us consider a system consisting of two subsystems, 1 and 2. The system is closed. The two subsystems, however, are open, and matter can be transferred between 1 and 2. The closed system is thermally insulated — an adiabatic system. The temperature of subsystem 1, T_1, is higher than the temperature of subsystem 2, T_2. The entropy production for three different irreversible processes taking place between the two subsystems will be studied:

1. Only heat is transferred from subsystem 1 to subsystem 2.
2. Both heat and matter are transferred from subsystem 1 to subsystem 2.
3. Heat, matter and electric charge are transferred from subsystem 1 to subsystem 2.

2.1.1 *Heat transfer only*

A system with only heat transfer can be exemplified as consisting of two pieces of metal at different absolute temperatures, T_1 and T_2, where $T_1 > T_2$. A small quantity of heat, dq, is allowed to pass from subsystem 1 to subsystem 2 in course of a short time interval, dt (Fig. 2.1). The quantity of heat is so small that changes in the temperatures of the subsystems can be disregarded.

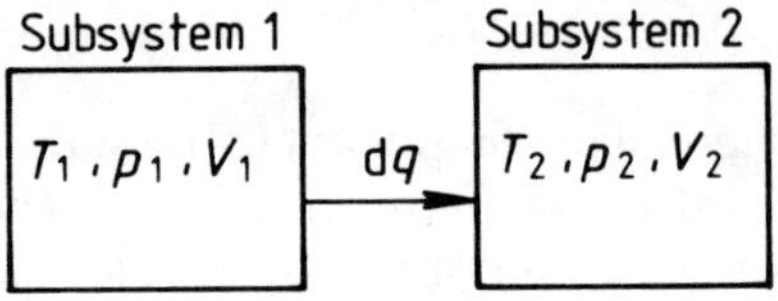

Figure 2.1. Heat transfer, dq.

Since there is no transfer of matter, each subsystem behaves as a closed system, and eq. (2.3) reduces to eq. (2.2). The transfer of dq leads to the following changes in the subsystems:

$$\text{Subsystem 1:}\quad dU_1 = T_1 dS_1 - p_1 dV_1 \qquad \text{Subsystem 2:}\quad dU_2 = T_2 dS_2 - p_2 dV_2 \tag{2.2}$$

or with the equations solved with respect to entropy change:

$$dS_1 = dU_1/T_1 + (p_1/T_1)dV_1 \qquad dS_2 = dU_2/T_2 + (p_2/T_2)dV_2$$

When pressure–volume work is the only work carried out, dU_1

and dU_2 can be expressed by the external variables, which are measurable quantities,

$$dU_1 = dq_1 - p_1 dV_1 \qquad dU_2 = dq_2 - p_2 dV_2$$

The total system is adiabatic. Thus we can write

$$dq = -dq_1 = dq_2$$

and the change in internal energy for the total system is

$$dU_1 + dU_2 = -p_1 dV_1 - p_2 dV_2$$

The production of entropy in the total system is

$$dS = dS_1 + dS_2$$

and we obtain

$$dS = -dq/T_1 + dq/T_2 = (1/T_2 - 1/T_1)dq = \Delta(1/T)dq \qquad (2.4)$$

The heat is transferred in the course of the time dt and the entropy production per unit time is

$$dS/dt = \Delta(1/T)\frac{dq}{dt} \qquad (2.5)$$

When both sides of eq. (2.5) are multiplied by T_2, the dissipated energy per unit time is obtained:

$$T_2\frac{dS}{dt} = T_2\Delta(1/T)\frac{dq}{dt} \qquad (2.6)$$

The dissipated energy, $T_2 dS$, is the part of the thermal energy received by subsystem 2 in course of the time dt, which could have been converted to work in a corresponding reversible process. If the heat, dq, leaving subsystem 1 had entered a reversible engine operating between T_1 and T_2, the efficiency, η, of such an engine would have been equal to $(T_1 - T_2)/T_1$, and according to the second law of thermodynamics, the work produced would have been

$$\eta dq = \frac{T_1 - T_2}{T_2} dq = T_2(1/T_2 - 1/T_1)\, dq = T_2 dS$$

The righthand side of eq. (2.6) may be considered as the product of two terms. The one term, $T_2\Delta(1/T)$, originates from the temperature difference between the two subsystems, the difference that creates the transport. This term is the *force*, X. The other term, dq/dt, gives the rate of transport, it is the *flux*, J. The product of the two terms defines the dissipation function:

$$\text{dissipation function: } X \cdot J = \text{force} \cdot \text{flux}$$

2.1.2 *Transfer of heat and matter*

A system with transfer of heat and matter can be exemplified by two gas containers at different absolute temperatures, T_1 and T_2, where $T_1 > T_2$ (Fig. 2.2). A small quantity of heat, $d\Phi$, and small quantities of the different gas components, dn_i, are allowed to pass from subsystem 1 to subsystem 2 in course of the time dt. *The total heat transferred*, $d\Phi$, is composed of a measurable heat and the enthalpy of the transferred matter. Since the total system is adiabatic, the total heat removed from subsystem 1 is equal to the total heat received by subsystem 2. The enthalpy of the transferred matter, however, may be different in the two subsystems, and thus the measureable heat change will be different in the two subsystems.

In this case eq. (2.3) must be used to describe the changes in the subsystems:

$$\begin{aligned} &\text{Subsystem 1} \\ dU_1 &= T_1 dS_1 - p_1 dV_1 + \Sigma_i \mu_{i,1} dn_{i,1} \\ &\text{Subsystem 2} \\ dU_2 &= T_2 dS_2 - p_2 dV_2 + \Sigma_i \mu_{i,2} dn_{i,2} \end{aligned} \tag{2.3}$$

or the equations solved with respect to entropy changes:

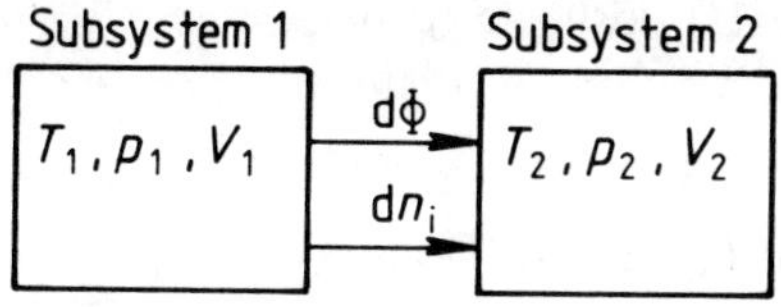

Figure 2.2. Transfer of heat, $d\Phi$, and matter, dn_i.

$$dS_1 = \frac{dU_1}{T_1} + \frac{p_1}{T_1} dV_1 - \frac{1}{T_1} \Sigma_i \mu_{i,1} dn_{i,1}$$

$$dS_2 = \frac{dU_2}{T_2} + \frac{p_2}{T_2} dV_2 - \frac{1}{T_2} \Sigma_i \mu_{i,2} dn_{i,2} \tag{2.7}$$

We can express dU_1 and dU_2 by external variables. When pressure–volume work is the only work carried out, we have

$$dU_1 = d\Phi_1 - p_1 dV_1 \qquad dU_2 = d\Phi_2 - p_2 dV_2 \tag{2.8}$$

and

$$dU_1 + dU_2 = - p_1 dV_1 - p_2 dV_2$$

Thus we have

$$d\Phi = - d\Phi_1 = d\Phi_2 \tag{2.9}$$

Since the two open systems together form a closed system, the amount of substance lost from subsystem 1 equals the amount of substance gained by subsystem 2 for each species,

$$dn_i = - dn_{i,1} = dn_{i,2} \tag{2.10}$$

Then, from eqs. (2.7), (2.8), (2.9) and (2.10)

$$dS_1 = - \frac{d\Phi}{T_1} + \frac{1}{T_1} \Sigma_i \mu_{i,1} dn_i \qquad dS_2 = \frac{d\Phi}{T_2} - \frac{1}{T_2} \Sigma_i \mu_{i,2} dn_i \tag{2.11}$$

The production of entropy in the total system is $dS = dS_1 + dS_2$, and therefore

$$dS = (1/T_2 - 1/T_1)d\Phi - \Sigma_i(\mu_{i,2}/T_2 - \mu_{i,1}/T_1)dn_i$$

or

$$dS = \Delta(1/T)d\Phi - \Sigma_i\Delta(\mu_i/T)dn_i \tag{2.12}$$

In order to obtain eq. (2.12) in a more useful form, we shall interpret $d\Phi$ in common thermodynamic terms, and develop a more suitable expression for $\Delta(\mu_i/T)$ of the second term. From eq. (2.11) we have

$$d\Phi = -T_1dS_1 + \Sigma_i\mu_{i,1}dn_{i,1} = T_2dS_2 + \Sigma_i\mu_{i,2}dn_{i,2} \tag{2.13}$$

The entropy of an open system can be expressed as a function of temperature, pressure and the amounts of substances, $S = f(T,p,n_i)$. The experiment can be arranged such that the pressure within each subsystem is kept constant. The change in entropy with temperature and amounts of substances for subsystem 1 can be written,

$$\begin{aligned} dS_1 &= (\partial S_1/\partial T)_{p,n_i}dT + \Sigma_i(\partial S_1/\partial n_{i,1})_{p,T,n_{j\neq i}}dn_{i,1} \\ &= (C_{p,1}/T_1)\,dT + \Sigma_i S_{i,1}dn_{i,1} \end{aligned} \tag{2.14}$$

and in a similar way for subsystem 2. The heat capacity of the system multiplied by the change in temperature is equal to the measurable heat absorbed by the system, $C_{p,1}dT = -dq_1$ where dq_1 is the heat *removed from* subsystem 1. Here it is assumed that dT is so small that changes in $C_{p,1}$ can be neglected. When the amounts of substances transferred from subsystem 1 to subsystem 2 are small, the change in partial molar entropy, $S_{i,1}$ can be neglected. Thus for subsystem 1,

$$dS_1 = -dq_1/T_1 - \Sigma_i S_{i,1}dn_i \tag{2.15}$$

This equation combines with eq. (2.11) for dS_1 to give

$$d\Phi = dq_1 + \Sigma_i(\mu_{i,1} + T_1S_{i,1})dn_i \tag{2.16}$$

or, since $H_i = \mu_i + TS_i$,

$$d\Phi = dq_1 + \Sigma_i H_{i,1}dn_i$$

In a similar way we can express $d\Phi$ by the heat *absorbed* by subsystem 2, dq_2, and the enthalpy added to subsystem 2 by the transfer of dn_i moles of substance:

$$d\Phi = dq_2 + \Sigma_i H_{i,2} dn_i \tag{2.17}$$

Comparing eqs. (2.16) and (2.17), it can be seen that the difference in the dq values corresponds to a difference in the H values:

$$dq_2 - dq_1 = \Sigma_i (H_{i,1} - H_{i,2}) dn_i$$

(For gases $H_{i,1} - H_{i,2} \approx 0$).

The enthalpies, H_i, do not have absolute values, and therefore $d\Phi$ does not have an absolute value. We shall choose subsystem 1 as the reference state, and replace $d\Phi$ in eq. (2.12) by the expression given in eq. (2.16). The second term in eq. (2.12), $\Sigma_i \Delta(\mu_i/T) dn_i$, can also be referred to subsystem 1 as the reference state. The chemical potentials are functions of temperature, pressure and composition, and for small differences in these parameters between the two subsystems, we have by the rules of derivation

$$\Delta(\mu_i/T) = \Delta(1/T)\mu_{i,1} + \frac{1}{T_1}(\partial\mu_{i,1}/\partial T)_{p,n_i} \Delta T + \frac{1}{T_1}\Delta\mu_{i,T} \tag{2.18}$$

where $\Delta\mu_{i,T}$ gives the *variation in* μ_i *with changes in composition and pressure.*

Since

$$(\partial\mu_{i,1}/\partial T)_{p,n_i} = -S_{i,1}$$

and

$$\Delta(1/T) = \left(\frac{1}{T+\Delta T} - \frac{1}{T}\right) \approx -\frac{\Delta T}{T^2}$$

and

$$T_1 \approx T$$

therefore,

$$\Sigma_i \Delta(\mu_i/T) dn_i = \Delta(1/T)\Sigma_i(\mu_{i,1} + TS_{i,1}) dn_i + \frac{1}{T}\Sigma_i \Delta\mu_{i,T} dn_i \tag{2.19}$$

Introducing eqs. (2.16) and (2.19) into (2.12) we obtain

$$dS = \Delta(1/T)dq_1 - \frac{1}{T}\Sigma_i \Delta\mu_{i,T} dn_i \tag{2.20}$$

2.1.3 *Transfer of heat, matter and electric charge*

A system with transfer of heat, matter and electric charge can be exemplified by an electrochemical cell (Fig. 2.3). The two half-cells are connected by a junction (e.g. a porous plug with electrolyte), which can transfer heat, mass and charge whilst maintaining a pressure difference. It is assumed that there is an approximately stationary state over the junction, i.e. that the entropy change in the junction is negligible.

The entropy production in the total system is the sum of the entropy changes in the subsystems. Since entropy is a state function, any convenient path from the initial state to the final state may be chosen for calculating the entropy production. No charge is accumulated in any of the subsystems, and the changes in each subsystem can be described by means of neutral components transferred from one subsystem to the other, even though the path of transport may be by means of ions.

Small quantities of heat, matter and electric charge are allowed to pass from subsystem 1 to subsystem 2. An electric work, dw_{el}, is carried out on the system, and we have

$$dU_1 + dU_2 = - p_1 dV_1 - p_2 dV_2 + dw_{el}$$

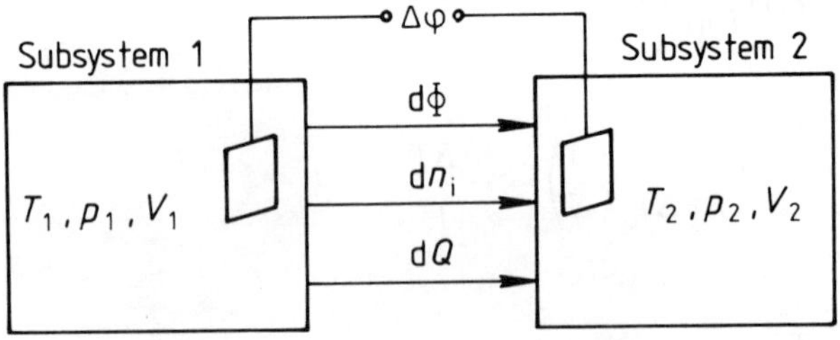

Figure 2.3. Transfer of heat, $d\Phi$, matter, dn_i, and electric charge, dQ. The potential of the electrochemical cell is $\Delta\varphi$. Reprinted with permission from *Electrochimica Acta*, **25**, Førland, T. and Ratkje, S.K., Entropy production by heat, mass, charge transfer and specific chemical reactions. Copyright (1980) Pergamon Journals Ltd.[1]

The electric work can be expressed as $-\Delta\varphi dQ$, the product of the potential and the transferred charge. (When there is a pressure difference between the two subsystems, the potential needs a correction for the volume changes of the electrodes. This will be discussed in Section 5.3.3). When the difference in temperature between the two subsystems is small, we can just add a term to eq. (2.20):

$$dS = \Delta(1/T)dq_1 - \frac{1}{T}\Sigma_i \Delta\mu_{i,T} dn_i - \frac{1}{T}\Delta\varphi dQ \qquad (2.21)$$

In this derivation of the entropy production, the electric potential entered the equation via external electric work as expressed in the first law. In a similar way other kinds of work can be included, e.g. gravitational work.

2.2 *Entropy Production in a Continuous Adiabatic System*

Many irreversible processes take place in a continuous system with gradual changes in properties. Measurements carried out on such systems are frequently of the integral type. The observed values may be considered as the sum of the contributions from local sections throughout the system. In order to relate integral properties to local parameters, there is a need for equations describing the differential changes.

We shall consider the entropy production by the transfer of heat, matter and charge across a membrane separating two mixtures, I and II. The mixtures are of different temperature, pressure and composition; each mixture, however, is uniform in these variables. The membrane allows a gradual change in the variables from mixture I to mixture II giving fluxes of heat and matter. With electrodes in the two mixtures and a closed electric circuit, there will also be a flux of charge, an electric current. It will be assumed that the process does not create macroscopic kinetic energy changes and that there is no transport of bulk solution through the membrane.

A frame of reference for the transport is needed in order to describe the local changes in composition (or any other intensive property). With no macroscopic kinetic energy changes, any frame of reference may be chosen — the container walls, the membrane matrix, the centre of gravity of the system or one of the components (see Agar,[2] Sundheim[3], and Kirkwood and coworkers[4] for further details). Here it is convenient to choose the membrane matrix as the frame of reference. Any change in the volume of the matrix by the transport is neglected. The membrane matrix reference is then the same as the container wall reference. (For systems without a membrane, it is convenient to choose the solvent as the frame of reference.)

For the calculation of local entropy production in the membrane, fluxes and changes in intensive variables will be described as taking place in small steps. Thus the membrane will be considered as composed of narrow sections or subsystems of uniform intensive variables (Fig. 2.4). These subsystems have fixed limits and do not change volume in the transport process. Any volume change will only take place in mixtures I and II. The total transfer process is equal to the sum of the transfers between all neighbouring pairs of subsystems.

The quantities transferred in one step are assumed small compared to the size of a subsystem. Thus, when considering one step, the changes in intensive variables of a subsystem can be

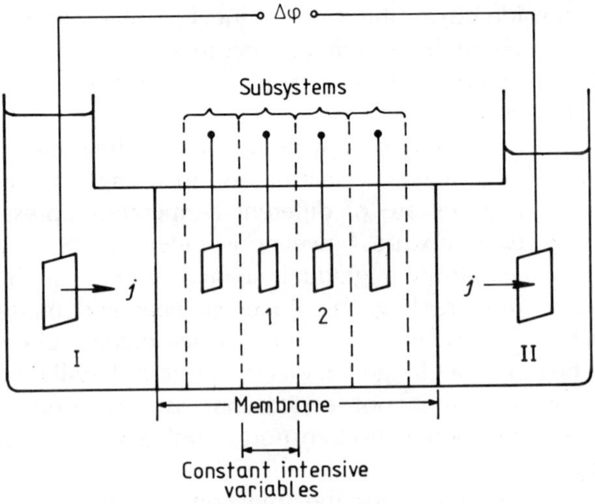

Figure 2.4. Transfer of heat, matter and charge in an adiabatic system composed of a series of subsystems. Reprinted with permission from *Electrochimica Acta* **25**, Førland, T. and Ratkje, S. K., Entropy production by heat, mass, charge transfer and specific chemical reactions. Copyright (1980) Pergamon Journals Ltd.[1]

neglected. The electric current flowing through all the subsystems is the same, i.e. there is no accumulation of charge anywhere in the system. Test electrodes of the same kind as the end electrodes are placed in the subsystems. They are only used for measuring local electric potentials. The currents passing through these electrodes in the stepwise charge transfer will cancel. For simplicity in calculation, we shall consider forces and fluxes in the x-direction only.

The entropy production by a one-step transfer from subsystem 1 to 2 (see Fig. 2.4), can be expressed in a similar way to eq. (2.21). If the length of each subsystem is dx, the entropy production per unit length in any region of the subsystem is given by

$$\frac{\mathrm{d}}{\mathrm{d}x}(\mathrm{d}S) = \frac{\mathrm{d}(1/T)}{\mathrm{d}x}\,\mathrm{d}q - \frac{1}{T}\Sigma_i(\mathrm{d}\mu_{i,T}/\mathrm{d}x)\mathrm{d}n_i - \frac{1}{T}(\mathrm{d}\varphi/\mathrm{d}x)\mathrm{d}Q \qquad (2.22)$$

The number of components is n. The membrane component, the frame of reference, is not included in the summation, only the n−1 components transported. In addition there is a heat term and an electric term; thus eq. (2.22) contains k = n+1 terms.

If the quantities $\mathrm{d}q$, $\mathrm{d}n_i$ and $\mathrm{d}Q$ pass a cross-section of unit area in course of the time $\mathrm{d}t$, the entropy production per unit volume and unit time, Θ, is obtained:

$$\Theta = \frac{\mathrm{d}}{\mathrm{d}x}\left(\frac{\mathrm{d}S}{\mathrm{d}t}\right) = \frac{\mathrm{d}(1/T)}{\mathrm{d}x}\left(\frac{\mathrm{d}q}{\mathrm{d}t}\right) - \frac{1}{T}\Sigma_i(\mathrm{d}\mu_{i,T}/\mathrm{d}x)\,(\mathrm{d}n_i/\mathrm{d}t)$$
$$- \frac{1}{T}(\mathrm{d}\varphi/\mathrm{d}x)(\mathrm{d}Q/\mathrm{d}t) \qquad (2.23)$$

where the fluxes are

$$\mathrm{d}q/\mathrm{d}t = J_q \;,\; \mathrm{d}n_i/\mathrm{d}t = J_i \;,\; \mathrm{d}Q/\mathrm{d}t = j \qquad (2.24)$$

When multiplying eq. (2.23) by T, the *dissipation function*, $T\Theta$, is obtained:

$$T\Theta = -\frac{\mathrm{d}\ln T}{\mathrm{d}x}J_q - \sum_{i=1}^{n-1}(\mathrm{d}\mu_{i,T}/\mathrm{d}x)J_i - \frac{\mathrm{d}\varphi}{\mathrm{d}x}\,j \qquad (2.25)$$

Defining the dissipation function as the sum of the force–flux products, we may write

$$T\Theta = \sum_{j=1}^{k} X_j J_j \tag{2.26}$$

where the summation includes heat, all transported components and electrical charge. The driving forces are

$$-\frac{d\ln T}{dx} = X_q \ , \quad -\frac{d\mu_{i,T}}{dx} = X_i \ , \quad -\frac{d\varphi}{dx} = X_j \tag{2.27}$$

One may refer forces and fluxes to the entropy production, Θ, instead of the dissipation function:

$$\Theta = \sum_{j=1}^{k} X_j J_j \tag{2.28}$$

The flux $J_q = dq/dt$ of eq. (2.24) may be replaced by $J_S = (1/T)\ dq/dt$, with the force $X_S = -\ d\ln T/dx$ (same as X_q in eq. (2.27)). The fluxes J_i and j are defined by eq. (2.24), giving the forces $X_i = -\ (1/T)\ d\mu_{i,T}/dx$ and $X_j = -\ (1/T)\ d\varphi/dx$ to replace the definitions given in eq. (2.27).

An advantage of the definitions in eq. (2.27) is that familiar expressions for the forces, such as derivatives of chemical and electrical potentials, are obtained while a division by temperature gives more unusual forces.

For transport in three dimensions eq. (2.25) is replaced by eq. (2.29):

$$T\Theta = -\ \nabla\ln T J_q - \Sigma_i \nabla\mu_{i,T} J_i\ -\ \nabla\varphi j \tag{2.29}$$

The corresponding equation in terms of entropy production is,

$$\Theta = -\ \nabla\ln T J_S - \Sigma_i \frac{1}{T}\nabla\mu_{i,T} J_i - \frac{1}{T}\nabla\varphi j \tag{2.30}$$

Equation (2.25) was developed assuming pressure differences. It is equally valid when there is no pressure difference. Then the force X_i varies only with composition. When there is no temperature difference, $X_q = 0$ and the first term of eq. (2.25) disappears. When there is no electric potential $X_j = 0$ and when we have an open circuit $j = 0$. In both cases the third term of eq. (2.25) disappears. For many applications only a simplified version of eq. (2.25) is needed.

In order to integrate eq. (2.25) over time and space, one must assume that the thermodynamic state functions have the same values as they would have in an equilibrium situation, *the assumption of local equilibrium*. The system is pictured as divided into several subsystems. *During* a one-step transfer intensive variables are assumed to be constant. *After* the transfer they are assumed to take the new values consistent with the change in composition.

In order to utilize the equations derived in this chapter, information must be obtained about the variables. Since the fluxes are created by the driving forces, the fluxes must be studied as functions of the forces, the flux equations. This will be dealt with in subsequent chapters.

2.3 *References*

1. Førland, T. and Ratkje, S. K., *Electrochim. Acta*, **25**, 157 (1980).
2. Agar, J. N., Thermogalvanic cells, in *Electrochemistry and Electrochemical Engineering* (ed. Delaney, P.), Interscience, New York, 1963.
3. Sundheim, B. R., Transport properties of liquid electrolytes, in *Fused Salts* (ed. Sundheim, B. R.), McGraw-Hill, New York, 1964.
4. Kirkwood, J. G., Baldwin, R. L., Dunlop, P. J., Gosting, L. J. and Kegeles, G., *J. Chem. Phys.*, **33**, 1505 (1960).

2.4 *Exercises*

2.1. *Thermodynamic properties*. (a) The chemical potential of component i, μ_i, can be expressed by the change in internal energy, U, with the change in the amount of the component. Similarly it can be expressed by the change in enthalpy, H, in Helmholtz energy, A, in Gibbs energy, G, and in entropy, S. Derive these expressions. (b) Show that the Gibbs–Duhem equation can be written, $\Sigma_i n_i \mathrm{d}\mu_i = V\mathrm{d}p - S\mathrm{d}T$.

2.2. *Dissipation function*. (a) Give a brief explanation of the concept 'dissipation function', and its relation to entropy production and to dissipated energy. (b) Explain the difference between the ways the gradients $\nabla\mu_i$ and $\nabla\varphi$ enter the equation for entropy production. (c) Why is the dissipation function $T\Theta \geq 0$? (d) What is the dissipation function for a metallic conductor at constant temperature when the current density is j? Use Ohm's law. (e) What is the dissipation function for the diffusion of one component in a system at constant temperature? Use Fick's law.

CHAPTER 3

The Flux Equations

In Chapter 1 some familiar examples of transport processes were mentioned, such as heat conduction, electric conduction and diffusion. The empirical laws for these processes express fluxes as linear functions of forces. Several phenomena of mixed transport, where fluxes are linear, homogeneous functions of forces, are also known. Linear relations are approximations that are valid when there is microscopic reversibility, i.e. local equilibrium. The concept of thermodynamic components was discussed. A system is described by the minimum number of neutral species needed to describe any part of the system.

The entropy production and the dissipation function for irreversible processes were discussed in Chapter 2. Equations were developed from fundamental thermodynamic principles, which are assumed valid when there is local equilibrium.

This chapter deals with the *flux equations*, when fluxes are *linear, homogeneous* functions of the forces, and the proportionality factors for the linear relations, the *phenomenological coefficients*. By choosing thermodynamic components, the independence of forces can be assured, and the relations between the coefficients under these conditions are studied.

3.1 *Flux Equations and Phenomenological Coefficients*

Any flux, J_i, can be expressed as a linear, homogeneous function of the forces in the following way:

$$J_i = \sum_{j=1}^{k} L_{ij} X_j; \quad (i = 1, \ldots, k) \tag{3.1}$$

Equation (3.1) can be written out in more detail:

$$J_i = L_{i1}X_1 + L_{i2}X_2 + \ldots + L_{ii}X_i + \ldots + L_{ij}X_j + \ldots + L_{ik}X_k$$

The forces, X_j, may be, for example, gradients in natural logarithms of temperature, in chemical potentials and in electric potential, with conjugate fluxes of heat (entropy), matter and electric charge. A complete set of equations contains the same number of fluxes and forces. The *phenomenological coefficients, L_{ij}, are independent of the forces.*

We may take as an example an isothermal system with fluxes of two components, J_1 and J_2, and of electric charge, $j = J_3$. The forces are $X_1 = -\nabla\mu_1$, $X_2 = -\nabla\mu_2$ and $X_3 = -\nabla\varphi$. The fluxes expressed by the forces are

$$\begin{aligned} J_1 &= -L_{11}\nabla\mu_1 - L_{12}\nabla\mu_2 - L_{13}\nabla\varphi \\ J_2 &= -L_{21}\nabla\mu_1 - L_{22}\nabla\mu_2 - L_{23}\nabla\varphi \\ j = J_3 &= -L_{31}\nabla\mu_1 - L_{32}\nabla\mu_2 - L_{33}\nabla\varphi \end{aligned} \tag{3.2}$$

The *direct coefficients*, L_{11}, L_{22} and L_{33}, relate *conjugate fluxes and forces*. The coefficients L_{11} and L_{22} are related to the diffusion coefficient in Fick's law, while the coefficient L_{33} may be considered as the electric conductivity of the system. The *cross coefficients* L_{12}, L_{13}, L_{21}, L_{23}, L_{31} and L_{32} represent cross phenomena. They refer to reactions of the components to forces that are acting directly on other components. Units must correspond in a flux equation. If fluxes, or forces, have different dimensions, this is reflected in the dimensions of the cross coefficients.

3.2 *The Onsager Reciprocal Relations*

According to Onsager[1, 2] the cross coefficients are related to each other in the following way:

$$L_{ij} = L_{ji}; \quad (i, j = 1, \ldots, k \text{ and } j \neq i) \tag{3.3}$$

Equation (3.3) is called the *Onsager Reciprocal Relations*. For our example, eq. (3.2), the relations are $L_{12} = L_{21}$, $L_{13} = L_{31}$ and $L_{23} = L_{32}$.

Onsager derived eq. (3.3) from the *principle of microscopic reversibility*, using statistical arguments. Chemists are familiar with the idea of chemical equilibrium as a situation of microscopic reversibility. Microscopic reversibility is also the situation for random motions. The laws determining the motions of single particles of a system are symmetric with respect to time. This means that a deviation from the average value of an intensive variable, e.g. a concentration, at a time t, followed by the deviation in a second concentration at the time $t + \Delta t$, is as frequent as the events in the opposite order; the deviation in the second concentration at time t is followed by the deviation in the first concentration at time $t + \Delta t$. This is often called *the time reversal invariance*. A derivation of the Onsager relations is given in Appendix 2.

The form of the flux equations is given by eq. (3.1), showing a linear relationship between forces and fluxes. Another relationship between forces and fluxes is given by eq. (2.26) for the dissipation function, $T\Theta = \Sigma_j X_j J_j$. Equations (3.1) and (2.26) are not sufficient to determine the values of the phenomenological coefficients.[3] A simple example of only two forces and fluxes will be used as an illustration:

$$\begin{aligned} J_1 &= L_{11}X_1 + (L_{12} + a)X_2 \\ J_2 &= (L_{21} - a)X_1 + L_{22}X_2 \end{aligned} \tag{3.4}$$

In eq. (3.4) the parameter a can take any value without altering the sum of the force – flux products, as can be seen from eq. (3.5):

$$\begin{aligned} T\Theta = L_{11}X_1^2 + L_{12}X_2X_1 & \quad \boxed{\begin{aligned} &+ aX_2X_1 \\ &- aX_1X_2 \end{aligned}} \\ + L_{22}X_2^2 + L_{21}X_1X_2 & \end{aligned} \tag{3.5}$$

When the Onsager relation, $L_{12} = L_{21}$, is valid, the parameter a can only be equal to zero. Thus, in addition to eqs. (3.1) and (2.26), eq. (3.3) must also be obeyed.

3.3 *Independence of Forces and Fluxes*

In order to examine more closely the condition of independent forces, we shall return to the local entropy production that was discussed in Section 2.2. The local entropy production per unit volume by the transfer of small quantities of heat, matter and electric charge over the distance dx in a cell of unit cross-section is given the notation (d/dx)(dS). When the quantities transferred are small, changes in the intensive variables are negligible, and the entropy production is completely determined by the quantities transferred, dq, dn_i and dQ:

$$\frac{d}{dx}(dS) = \frac{d}{dx}\left(\frac{\partial S}{\partial q}\right)dq + \Sigma_i \frac{d}{dx}\left(\frac{\partial S}{\partial n_i}\right)dn_i + \frac{d}{dx}\left(\frac{\partial S}{\partial Q}\right)dQ \qquad (3.6)$$

This equation may be compared to eq. (2.22),

$$\frac{d}{dx}(dS) = \frac{d(1/T)}{dx}dq - \frac{1}{T}\Sigma_i(d\mu_{i,T}/dx)dn_i - \frac{1}{T}(d\varphi/dx)dQ \qquad (2.22)$$

Each term on the right-hand side of eq. (3.6) can be recognized as a product of a force and a quantity transferred — or, with the more familiar definition of forces given in eq. (2.27), 1/T times a product of a force and a quantity transferred.

The first term on the right-hand side gives the entropy production by transfer of heat. This can be obtained independently of the other contributions to entropy production by performing an experiment where all dn_i = 0 and dQ = 0, or, alternatively, where $d\mu_{i,T}/dx = 0$ and $d\varphi/dx = 0$:

$$\frac{d}{dx}\left(\frac{\partial S}{\partial q}\right) = \frac{d(1/T)}{dx} = -\frac{1}{T}\frac{d\ln T}{dx} = \frac{1}{T}X_q \qquad (3.7)$$

In a similar way the entropy production by charge transfer can be obtained independently of other contributions by arranging an experiment such that dq = 0 and all dn_i = 0, or, alternatively, $d\ln T/dx = 0$ and $d\mu_{i,T}/dx = 0$:

$$\frac{d}{dx}\left(\frac{\partial S}{\partial Q}\right) = \frac{1}{T}\frac{d}{dx}\left(\frac{\partial w_{el}}{\partial Q}\right) = -\frac{1}{T}\frac{d\varphi}{dx} = \frac{1}{T}X_j \qquad (3.8)$$

(An experiment with all dn_i = 0 may be performed by setting up concentration gradients such that transport by diffusion and charge transfer just balance each other).

The terms of eq. (3.6) containing dn_i can be studied experimentally when keeping dq = 0, $dn_{j\neq i} = 0$ and dQ = 0, or $d\ln T/dx = 0$, $d\mu_{j\neq i,T} = 0$ and $d\varphi/dx = 0$. Thus:

$$\frac{d}{dx}\left(\frac{\partial S}{\partial n_i}\right) = -\frac{1}{T}\frac{d\mu_{i,T}}{dx} = \frac{1}{T}X_i \tag{3.9}$$

for each *independent* force, i.e. for each *independent* change in chemical potential.

In order to make sure that the changes in chemical potentials are independent, the composition of the mixture must be described by the minimum number of pure components in the thermodynamic sense. This was discussed in Section 1.3. We must also have a defined frame of reference for the transports.

The composition of a solution may conveniently be described by concentrations of the components, e.g. the mole fractions,

$$x_i = n_i/\Sigma_i n_i \tag{3.10}$$

where

$$\Sigma_i x_i = 1 \quad \text{and} \quad \Sigma_i dx_i = 0 \tag{3.11}$$

Thus, if there are n components in a solution (a one-phase system), there are only n − 1 independent mole fractions. Correspondingly the chemical potentials are interrelated by the *Gibbs–Duhem equation*,

$$\Sigma_i x_i d\mu_{i,T} = 0; \quad (p,T = \text{const}) \tag{3.12}$$

At constant pressure all, except one, $\mu_{i,T}$ can be varied independently. One of the components is chosen as the frame of reference, and all changes refer to this frame. This means that the contributions to the entropy production given by eq. (3.9) are summed up for n − 1 components, and thus the forces X_i are independent. For transport in a solution the solvent may be chosen as the frame of reference. In an aqueous solution, for instance, the transport of salts is considered relative to water. In Section 2.2 a system was considered where the membrane matrix was chosen as the frame of reference. The membrane as a component and as the frame of reference, will be discussed in Chapter 5.

The time derivative of eq. (2.22) gives eq. (2.23), the entropy production per unit volume and unit time, Θ, and the fluxes created by the forces are given by eq. (2.24). When one component has been chosen as the frame of reference, only n − 1 material fluxes (in addition to J_q and j) are considered. Under certain conditions some of the fluxes may be interdependent (e.g. ionic fluxes in membranes). It can be shown that with independent forces, the Onsager reciprocal relations, eq. (3.3), are valid, even if the fluxes are interdependent.[4, 5]

As Onsager[1] pointed out, the relations are not valid for magnetic fields and Coriolis forces. Casimir[6] developed modified relations for these types of forces. They will not be discussed in this book.

3.4 *Restrictions on the Values of the Phenomenological Coefficients*

When the Onsager reciprocal relations, eq. (3.3), are obeyed, the set of phenomenological coefficients forms a symmetric matrix. In addition to the Onsager relations, there are some more restrictions on the phenomenological coefficients.

Since an irreversible process always leads to an increase in entropy, the dissipation function must have a positive value:

$$T\Theta = \Sigma_i X_i J_i \geq 0 \tag{3.13}$$

This means that the direct coefficients, L_{ii}, coupling the fluxes to their main driving forces, must be positive:

$$L_{ii} > 0; \quad (i = 1,2,3,\ldots) \tag{3.14}$$

The simplest example is a system of only one force and one flux, $J_1 = L_{11}X_1$. The equation expresses that a positive L_{11} gives a flux in the direction of the force.

For a system of two forces and fluxes:

$$J_1 = L_{11}X_1 + L_{12}X_2$$
$$J_2 = L_{21}X_1 + L_{22}X_2$$

one of the forces, e.g. X_2, may be eliminated from the equations. Then

$$J_1 = (L_{11} - L_{12}L_{21}/L_{22})X_1 + (L_{12}/L_{22})J_2$$

or

$$J_1 = l_{11}X_1 + (L_{12}/L_{22})J_2 \tag{3.15}$$

where

$$l_{11} = L_{11} - L_{12}L_{21}/L_{22} \tag{3.16}$$

The coefficient l_{11} is a direct coefficient, coupling the flux J_1 to its main driving force, X_1. With independent forces we can choose a value of X_2 that gives $J_2 = 0$. This reduces the system to one flux, $J_1 = l_{11}X_1$, and the direct coefficient l_{11} cannot be negative. If the fluxes are linearly dependent ($J_1 = \alpha J_2$) we have $l_{11} = 0$ and the coupling coefficient $\alpha = L_{12}/L_{22}$. With independent forces and fluxes

$$L_{11}L_{22} \geq L_{12}^2 \tag{3.17}$$

since $L_{12} = L_{21}$.

Chapters 4 and 5 show how we utilize the method of eliminating one of the forces from a set of flux equations to separate diffusion and migration by an electric current.

Let us consider a system of several forces and fluxes. In the general case the problem can always be reduced to a situation with only two fluxes and forces, by choosing all forces except X_i and X_j equal to zero. When all forces are independent, and J_i and J_j are independent of each other:

$$L_{ii}L_{jj} \geq L_{ij}^2 \tag{3.18}$$

Equation (3.1) expresses that fluxes may be written as linear homogeneous functions of the forces. Diffusion fluxes, heat fluxes and electric currents are vectors, and so are the forces, the gradients in natural logarithm of temperature, in chemical potentials and in electric potential. The positive values of the direct coefficients, L_{ii}, express that the forces direct the fluxes. In an isotropic system, where the thermodynamic properties are the same throughout the system, there would be no vectorial force and therefore no directed vectorial flux. There are, however, irreversible processes that do not involve directed diffusion. A chemical reaction taking place in an isotropic fluid is one example. In this case the forces are scalar, and the fluxes can be defined as the rate of increase in the amount of reaction products. The fluxes, as well as the forces, are scalar.

In an isotropic system there may not be any coupling between scalar and vectorial quantities. This is often called the *Curie principle*[4, 7]. In the case where J_1 and X_1 are scalars while J_2 and X_2 are vectors, the cross coefficients must be equal to zero,

$$L_{12} = L_{21} = 0 \tag{3.19}$$

An important consequence of the Curie principle is that a spontaneous chemical reaction in an isotropic fluid cannot be utilized for producing mechanical work. In many anisotropic biological systems, however, a chemical reaction can cause a flux of components across a membrane (active transport). Also, a chemical reaction can produce mechanical work in a muscle. This will be discussed in Section 10.1.

3.5 *Approximate Flux Equations for a System in a Stationary State*

Many irreversible processes approach stationary state with time. For systems in a stationary state with transfers over a temperature difference of only a few degrees, forces can be expressed by differences across the system instead of gradients.

The transports through the system may be pictured as transports between two reservoirs. Both reservoirs are so large that intensive variables undergo only negligible changes by the transfers between them (see Fig. 3.1). The system is small compared to the reservoirs,

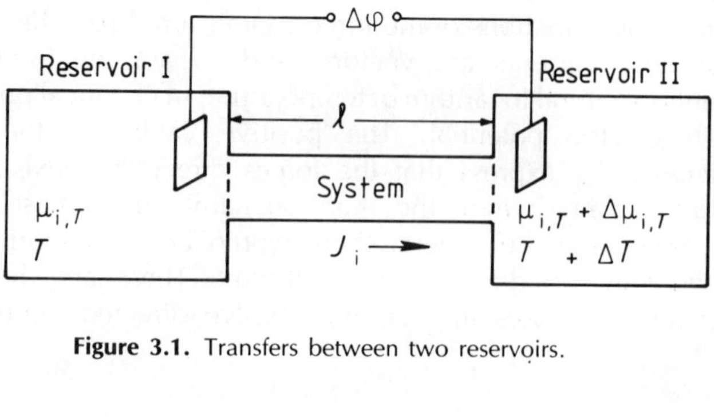

Figure 3.1. Transfers between two reservoirs.

and a stationary state is established after some time. In the stationary state the composition and gradients in the system do not change with time. In this case we can consider the system plus the two reservoirs as one discontinuous adiabatic system, such as the one treated in Section 2.1.3. The entropy production for this system is given by an equation similar to eq. (2.21):

$$dS = \Delta(1/T)dq - \frac{1}{T}\Sigma_i \Delta\mu_{i,T} dn_i - (\Delta\varphi/T)dQ \tag{3.20}$$

and the corresponding dissipated energy is given by

$$TdS = -\Delta\ln T dq - \Sigma_i \Delta\mu_{i,T} dn_i - \Delta\varphi dQ \tag{3.21}$$

The cross-section may vary through the system, and it is convenient to define fluxes as flows of heat, mass and charge per unit time *over the whole cross-section* of the system between the two reservoirs (not over unit area as in eq. (2.24)). The fluxes of heat and matter over the whole cross-section will be given the same symbols as before, J_q and J_i, while the electric current is given the symbol I to distinguish it from current density. The corresponding forces are $-\Delta\ln T$, $-\Delta\mu_{i,T}$ and $-\Delta\varphi$. The dissipated energy per time unit for the total system (including also the reservoirs) is

$$\frac{TdS}{dt} = -\Delta\ln T J_q - \Sigma_i \Delta\mu_{i,T} J_i - \Delta\varphi I \tag{3.22}$$

The corresponding flux equations are

$$J_j = -\bar{L}_{j1}\Delta\ln T - \sum_{i=2}^{k-1} \bar{L}_{ji}\Delta\mu_{i,T} - \bar{L}_{jk}\Delta\varphi; \quad (j = 1, \ldots, k) \tag{3.23}$$

In eq. (3.23) J_1 is the heat flux J_q, and $J_2, \ldots, J_{k-1}$ are the material fluxes, over the whole cross-section, while J_k is the current, I. The coefficients, $\bar{L}_{ij}$ are also for the whole cross-section. They have average values over the path of transport. The Onsager reciprocal relations are assumed to be valid for the average coefficients:

$$\bar{L}_{ij} = \bar{L}_{ji} \tag{3.24}$$

There are restrictions on the average coefficients similar to those discussed for the local coefficients in Section 3.4.

In some cases it may be preferable to define the forces as average forces over the length of the system, ℓ, $-\Delta\ln T/\ell$, $-\Delta\mu_{i,T}/\ell$ and $-\Delta\varphi/\ell$. If we divide by ℓ on both sides of eq. (3.23), the equation will represent the dissipated energy per unit time per unit length. The corresponding flux equations are:

$$J_j = -(\bar{L}_{j1}\ell)(\Delta\ln T/\ell) - \sum_{i=2}^{k-1} (\bar{L}_{ji}\ell)(\Delta\mu_{i,T}/\ell) - (\bar{L}_{jk}\ell)(\Delta\varphi/\ell);$$
$$(j = 1, \ldots, k) \tag{3.25}$$

where $(\bar{L}_{ij}\ell)$ represent the new coefficients.

The term *dissipation function*, $T\Theta$, is reserved for the dissipated energy per unit time, *per unit volume*:

$$T\Theta = \frac{1}{V} \cdot \frac{TdS}{dt} \tag{3.26}$$

When there is a large temperature difference in a system, it may be advantageous to refer forces and fluxes to the entropy production instead of the dissipated energy.

Flux equations are general transport equations. In the following chapters the equations will be applied to specific cases of transport. The physical interpretation of the coefficients will be studied.

3.6 *References*

1. Onsager, L., *Phys. Rev.*, **37**, 405 (1931).
2. Onsager, L., *Phys. Rev.*, **38**, 2265 (1931).
3. Coleman, B. D. and Truesdell, C., *J. Chem. Phys.*, **33**, 28 (1960).
4. de Groot, S. R. and Mazur, P., *Non-Equilibrium Thermodynamics*, North-Holland, Amsterdam, 1962.
5. Haase, R., *Thermodynamics of Irreversible Processes*, Addison-Wesley, New York, 1969.
6. Casimir, H. B. G., *Rev. Mod. Phys.*, **17**, 343 (1945).
7. Prigogine, I., *Physica*, **15**, 272 (1949); *J. Phys. Chem.*, **55**, 765 (1951).

3.7 *Exercises*

3.1. *Coefficients in flux equations*. The flux equations for transport of two solutes (in H_2O) can be written, $J_i = \Sigma_{j=1}^{2} L_{ij}X_j$, (i = 1, 2.)

(a) Write out the two equations in detail. (b) Why must L_{11} and L_{22} always be positive? (c) The coefficient L_{12} may be positive or negative. What would be the sign for L_{12} in flux equations for the transports discussed in Exercise 1.3? (d) What is the relation between L_{12} and L_{21}? (e) Derive the relation $L_{11}L_{22} \geq L_{12}^2$. (f) Under what conditions do we have $L_{12} = 0$? Under what conditions do we have $L_{11}L_{22} - L_{12}^2 = 0$?

3.2. *Independence of forces.* (a) A tube contains a mixture of the neutral species A, B and C. The composition is different in the two ends of the tube. At constant temperature and pressure a diffusion takes place, levelling out the difference in composition. How many independent forces for diffusion are there? (b) The tube contains the species Na^+, Cl^- and H_2O. How many independent forces for diffusion are there in this case? (c) The tube contains Na^+, K^+, Cl^- and H_2O. How many independent forces for diffusion are there now?

3.3. *Flux equations and the dissipation function.* Consider the three-component system discussed in Exercise 3.1. (a) Express the dissipation function by X_1, J_2 and coefficients (eq. (3.15) is helpful). Collect the result in three terms, the first term does not contain J_2, the second term does not contain X_1, while the third term contains X_1J_2. (b) Assume that $X_1 = -\nabla\mu_1$ and that $J_2 = j$. Give a physical interpretation of terms one and two, and give the sign (plus or minus) for each term. Use the Onsager reciprocal relations and find the value of the third term.

CHAPTER 4

Transport Processes in Concentration Cells

In the first three chapters the foundation of irreversible thermodynamics was laid. The basic conditions and assumptions were studied in Chapter 1, while Chapter 2 interpreted the entropy production in terms of thermodynamic state functions. In Chapter 3 the flux equations were scrutinized and relations between coefficients were established.

In this chapter we shall apply the flux equations to some simple isothermal and isobaric systems. We shall mainly deal with electrochemical cells with liquid junctions. The correlation between measurable parameters in a transport process will be studied. These correlations give us choices as to what kind of measurements to carry out to obtain experimental parameters. It will be shown how the contributions to the emf of a cell from changes in the liquid junction and at the electrode can be calculated from measurable quantities.

Concentration cells with *electrodes reversible to one ion present in the electrolytes* will be studied. The AgCl|Ag electrode is reversible to Cl^-, and the $H^+|H_2$ electrode is reversible to H^+. The two electrolytes of different concentration are connected by a *liquid junction*, where the concentrations are changing gradually from the constant concentrations of electrolyte I to the constant concentrations of electrolyte II.

A liquid junction can be obtained in different ways — by means of a salt bridge, a porous wall or plug, or any other device that prevents convective mixing of the two solutions, without hindering diffusion and electric transport. The different devices will not be discussed. An ideal liquid junction with no preferred adsorption or any other hindrance selective to the components will be assumed.

An interpretation of Fick's law of diffusion will be given and Fick's diffusion coefficient will be related to the phenomenological coefficients of the flux equations. The Nernst–Planck equation of electrochemical diffusion will also be interpreted in terms of the phenomenological coefficients. The validity of the law and the assumptions behind it in the more complex situations (more than one salt in solution) are revealed.

4.1 *Flux Equations — Determination of Coefficients*

4.1.1 *An HCl–H_2O concentration cell with AgCl|Ag Electrodes*

As a simple example consider the following concentration cell at constant temperature and pressure:

$$Ag(s)|AgCl(s)|HCl(aq,c_I)||HCl(aq,c_{II})|AgCl(s)|Ag(s)$$

The two electrodes are both AgCl|Ag electrodes. The two aqueous solutions of HCl, separated by a liquid junction, are of different concentration (see Fig. 4.1).

With a closed circuit, an electric current will pass through the cell, chemical reactions will take place at the electrodes and a transport of matter will take place in the electrolyte. We shall consider these changes, when one mole of electrons passes through

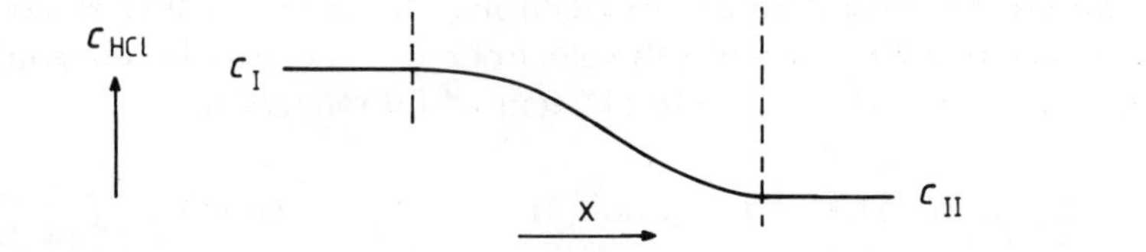

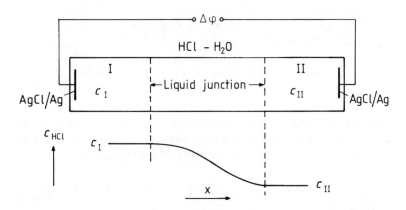

Figure 4.1. An HCl–H_2O concentration cell with AgCl|Ag electrodes.

the outer circuit from left to right, i.e. one faraday of positive charge passes through the *inner circuit* from left to right.

At the electrodes the chemical reactions are the following:

$$\text{left-hand-side electrode: } Ag(s) + Cl^- \rightarrow AgCl(s) + e^-$$
$$\text{right-hand-side electrode: } AgCl(s) + e^- \rightarrow Ag(s) + Cl^-$$

One mole Ag is transferred from the left-hand-side electrode to the right-hand-side one, while one mole AgCl is transferred in the opposite direction. One mole Cl^- is removed from the electrolyte at the electrode on the left-hand side, while one mol Cl^- is added to the electrolyte at the electrode on the right-hand side.

The aqueous solution is always electrically neutral, and the transfer of one mole Cl^- from left to right is compensated for by the migration of Cl^- to the left and of H^+ to the right in the electrolyte. The ions Cl^- and H^+ are the current carriers in the electrolyte. The fraction of the current carried by Cl^- is t_{Cl^-}, *the transference number* of Cl^-, while the fraction of the current carried by H^+ is t_{H^+}, *the transference number* of H^+. The transference numbers vary slightly with the concentration of HCl, but as Cl^- and H^+ are the only current carriers, their sum is always equal to unity:

$$t_{Cl^-} + t_{H^+} = 1 \tag{4.1}$$

The total changes caused by the transfer of charge in electrolyte I on the lefthand side are thus: one mole Cl^- removed at the electrode, $t_{Cl^-,I}$ mole Cl^- gained by migration and $t_{H^+,I}$ mole H^+ removed by migration:

$$\underbrace{(-1 + t_{Cl^-,I})}_{-t_{H^+,I}} \text{ mole } Cl^- - t_{H^+,I} \text{ mole } H^+ = -\, t_{H^+,I} \text{ mole HCl} \tag{4.2}$$

Similarly the total changes in electrolyte II on the righthand side are: one mole Cl^- gained at the electrode, $t_{Cl^-,II}$ mole Cl^- removed by migration and $t_{H^+,II}$ mole H^+ gained by migration:

$$\underbrace{(1 - t_{Cl^-,II})}_{t_{H^+,II}} \text{ mole } Cl^- + t_{H^+,II} \text{ mole } H^+ = +\, t_{H^+,II} \text{ mole HCl} \tag{4.3}$$

The difference, $(t_{H^+,I} - t_{H^+,II})$ mole HCl, is retained in the liquid junction.

As was discussed in Section 3.3, only the minimum number of components in the thermodynamic sense will be considered. The electrolytes consist of the two components HCl and H_2O. Water may be chosen as the frame of reference. For an irreversible process the changes taking place locally, anywhere in the electrolytes, can be expressed by the two fluxes J_{HCl} and j. The forces bringing about the changes are $X_{HCl} = -\nabla\mu_{HCl}$ and $X_j = -\nabla\varphi$. We are now dealing with an isothermal system, where the force $X_q = -\nabla \ln T = 0$, and heat flux is not included in a complete set of flux equations. The transfer of Ag and AgCl between the electrodes does not represent any change in chemical potential, the force being equal to zero. The flux equations expressing the changes are

$$J_{HCl} = -L_{11}\nabla\mu_{HCl} - L_{12}\nabla\varphi \tag{4.4a}$$

$$j = -L_{21}\nabla\mu_{HCl} - L_{22}\nabla\varphi \tag{4.4b}$$

It is convenient to use the same units for both fluxes. When J_{HCl} has the dimension mole m^{-2} s^{-1}, the corresponding dimension for the current density, j, is faraday m^{-2} s^{-1}, where the unit faraday means one mole elementary electric charges. If the same current density is measured as j_A A $m^{-2} = j_A$ C s^{-1} m^{-2}, we will have the ratio $j_A/j = 96{,}500$ C faraday^{-1} $= F$, or $j = j_A/F$, where F is the Faraday constant. Similarly, when $\nabla\mu_{HCl}$ has the dimension J mol^{-1} m^{-1}, the corresponding dimension for $\nabla\varphi$ is J faraday^{-1} m^{-1}. If the electric potential is measured in volts, the corresponding voltage gradient is ∇E V $m^{-1} = \nabla E$ J C^{-1} m^{-1}. This gives the ratio $\nabla\varphi/\nabla E = 96{,}500$ C faraday^{-1} $= F$, or $\nabla\varphi = \nabla EF$. A phenomenological coefficient, such as L_{22}, will have the dimension faraday m^{-2} s^{-1}/(J faraday^{-1} m^{-1}) = faraday^2 J^{-1} m^{-1} s^{-1}.

In Section 3.5 the symbols J_q, J_i and I were used for the fluxes of heat, mass and charge *over the whole cross-section*. Here I is the current and has the dimension faraday s^{-1}. The corresponding force, $\Delta\varphi$, has the dimension J faraday^{-1}, and the average coefficient $\bar{L}_{kk}$ has the dimension faraday s^{-1}/(J faraday^{-1}) = faraday^2 J^{-1} s^{-1}.

The Onsager reciprocal relations require that $L_{12} = L_{21}$. Further, L_{11} and L_{22} have positive values, and $L_{11}L_{22} \geq L_{12}^2$ (see eqs. (3.3), (3.14) and (3.18)). The phenomenological coefficients depend on local concentrations, but they are independent of gradients. Under specified conditions the values of the coefficients may be determined experimentally.

The cell can be arranged such that gradients occur only in the direction of the x-axis between the electrodes. Then $X_{HCl} = - d\mu_{HCl}/dx$ and $X_j = - d\varphi/dx$. Both gradients are well defined anywhere in the cell as the limiting value of observable quantities, e.g. $d\mu_{HCl}/dx = \lim_{\Delta x \to 0} \Delta\mu_{HCl}/\Delta x$. The concept of the unmeasurable electric potential between electrode and solution is avoided. The two forces may be varied independently, X_{HCl} by changing concentrations and X_j by changing the applied potential.

Determination of L_{22} at constant composition. When $d\mu_{HCl}/dx = 0$, the coefficient L_{22} is obtained from eq. (4.4b) as

$$L_{22} = - \frac{j}{d\varphi/dx} \tag{4.5}$$

The coefficient L_{22} is related to the *electric conductivity*, κ, at the given composition of the electrolyte. The conductivity can be obtained by standard methods of conductivity measurements. When using the SI units ampere and volt, the electric conductivity is given by $\kappa = - j_A/\nabla E$ ohm^{-1} m^{-1}. With the relations $j_A/j = F$ and $\nabla\varphi/\nabla E = F$, we obtain

$$L_{22} = \kappa/F^2 \tag{4.6}$$

Determination of L_{12} at constant composition. When $d\mu_{HCl}/dx = 0$, the following is obtained from eq. (4.4):

$$J_{HCl}/j = L_{12}/L_{22} = t_{HCl} \tag{4.7}$$

Here J_{HCl}/j is the number of moles HCl transferred with one faraday electric charge. It may be named the *transference coefficient* of HCl, t_{HCl}. With AgCl|Ag electrodes it is equal to the transference number of H^+, $t_{HCl} = t_{H^+}$ (compare eqs. (4.2) and (4.3)). A transference coefficient is not always equal to a transference number, although related to it, as will be seen from subsequent examples. While transference numbers must have positive values, transference coefficients may be negative. Also, the sum of the transference coefficients may be different from unity.

There are several different methods for the determination of transference numbers. We shall consider the *Hittorf method*.

Consider a cell as shown schematically in Fig. 4.2. At the start of the experiment the electrolyte is of constant composition, $d\mu_{HCl}/dx = 0$. If electrons are passed from left to right in the outer circuit, the current will be carried through the solution by Cl^- ions migrating from right to left and by H^+ ions migrating in the

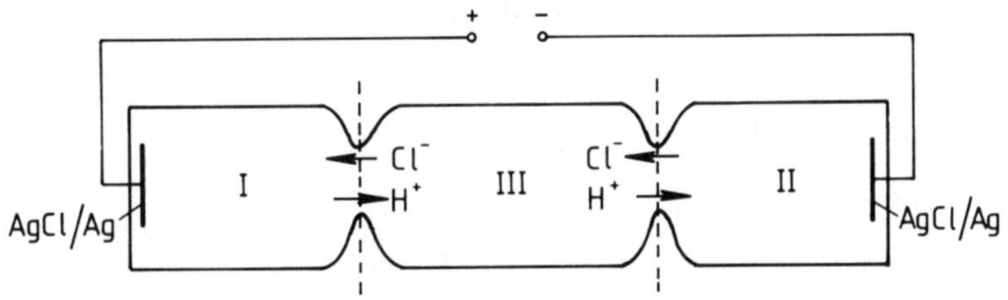

Figure 4.2. A Hittorf cell.

opposite direction. This leads to changes in the cell. When passing one faraday of electric current, electrolyte I will loose t_{H^+} mole HCl while electrolyte II will gain t_{H^+} mole HCl (compare eqs. (4.2) and (4.3)). The transports at the two junctions I – III and III – II are identical, and there will be no change in electrolyte III. After passing a known current, the composition of each electrolyte is determined analytically. The number of moles HCl gained per faraday in electrolyte II (or lost in electrolyte I) is equal to the transference number t_{H^+}, which gives L_{12} through eq. (4.7) when L_{22} is known. For more details about the Hittorf method, and for other methods of measuring transference numbers, see, for example, refs. 1, 2.

Determination of L_{21} at zero current density. Since $L_{12} = L_{21}$ by the Onsager relations, the cross coefficient may also be expressed by another relation than eq. (4.7). A cell with a concentration gradient $d\mu_{HCl}/dx \neq 0$ is used. The emf of the cell is counterbalanced by means of a potentiometer so that $j \approx 0$ in the closed circuit. Thus by eq. (4.4b):

$$-\,d\varphi/d\mu_{HCl} = L_{21}/L_{22} = t_{HCl} \tag{4.8}$$

By measuring the emf across a cell with a known difference in concentration between the electrolytes an average value for t_{HCl} over the concentration interval can be found, $\bar{t}_{HCl} = -\,\Delta\varphi/\Delta\mu_{HCl}$. By making the concentration interval small, the value of t_{HCl} at one concentration is approached.

Determination of L_{11} at zero current density. Elimination of $\nabla\varphi$ in eqs. (4.4) gives

$$J_{HCl} = -\,(L_{11} - L_{12}L_{21}/L_{22})\nabla\mu_{HCl} + (L_{12}/L_{22})j$$

or

$$J_{HCl} = -\,l_{11}\nabla\mu_{HCl} + t_{HCl}\,j \tag{4.9}$$

where

$$l_{11} = L_{11} - L_{12}L_{21}/L_{22} \tag{4.10}$$

The l_{11} is called a *diffusion coefficient*. The value of l_{11} may be determined by a pure diffusion experiment, $j = 0$. By measuring the flux of HCl in a cell with a known difference in concentration across the cell, an average value, $\bar{l}_{11}$, over the concentration interval can be found:

$$\bar{l}_{11} = -\,J_{HCl}/\Delta\mu_{HCl} \tag{4.11}$$

By making the concentration interval small, the value of l_{11} at one concentration is approached. When L_{12} and L_{22} are known, L_{11} can be calculated by the use of eq. (4.10).

The relation between L_{11} and L_{12}. When developing the equation for the entropy production (eq. (2.23)) the transfer of heat, matter and charge through a series of subsystems with test electrodes was studied (see Fig. 2.4). We shall consider two such adjacent subsystems (see Fig. 4.3). In these subsystems the distance between the AgCl|Ag electrodes is dx. The difference in chemical potential between the electrolytes is $\mu_{HCl,2} - \mu_{HCl,1} = d\mu_{HCl}$. The electrodes are short-circuited, and in the ideal case there is no resistance to an electric current passing through the outer circuit,

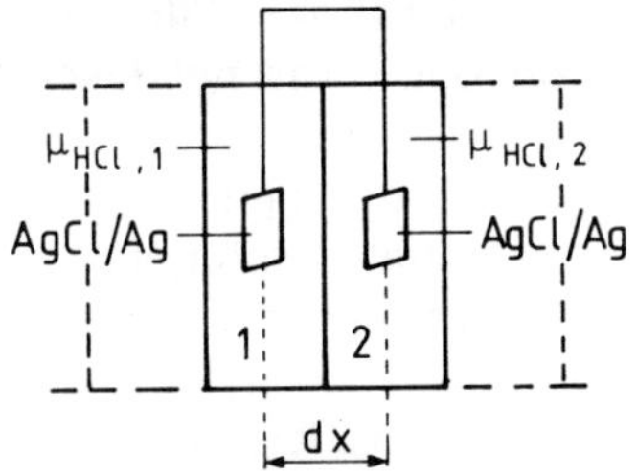

Figure 4.3. Two adjacent subsystems with short-circuited electrodes (cf. Fig. 2.4.).

and the potential equals zero, $d\varphi = 0$. Under these conditions eqs. (4.4) reduce to

$$J_{HCl} = -L_{11}\,d\mu_{HCl}/dx$$
$$j = -L_{12}\,d\mu_{HCl}/dx$$

or

$$J_{HCl}/j = L_{11}/L_{12} \tag{4.12}$$

The H^+ ions are transferred from left to right by migration through the electrolyte. For the Cl^- ions there are two possible routes of transport. They may migrate through the electrolyte or be transferred by a discharge of Cl^- ions at the left-hand-side electrode simultaneous with a production of Cl^- ions at the right-hand-side electrode. The ratio between the amounts taking either route is determined by the resistance to the movements by the two processes. In our ideal situation the resistance in the metal wires connecting the electrodes is equal to zero, and there is no overvoltage on the electrodes. In that case Cl^- ions will be transferred only via the electrodes:

$$(J_{HCl}/j)_{\nabla\varphi = 0} = 1 \tag{4.13}$$

and by eq. (4.12) we obtain

$$L_{11} = L_{21} \tag{4.14}$$

Since phenomenological coefficients are independent of gradients, the relation given by eq. (4.14) is still valid when $d\varphi \neq 0$.

When developing eq. (4.14) any diffusion in the electrolyte in the form of neutral ion pairs was not taken into account. This would give $L_{11} > L_{21}$. In a very dilute solution this diffusion of neutral ion pairs is probably negligibly small, while it may be substantial in more concentrated solutions.

Equation (4.14) was developed for a system with *only one salt* in the electrolyte solution. When there are two or more salts in the solution, we have to consider the cross phenomena — i.e. that the diffusion of one component is affected by the gradient in chemical potential for the other components.

4.1.2 *An HCl–H_2O concentration cell with $H^+|H_2$ Electrodes*

The cell is similar to the cell described in Section 4.1.1, except for the electrodes:

$$(\text{Pt})\text{H}_2(\text{g},1\ \text{atm})|\text{HCl}(\text{aq},c_{\text{I}})||\text{HCl}(\text{aq},c_{\text{II}})|\text{H}_2(\text{g},1\ \text{atm})\ (\text{Pt})$$

Both electrodes are $H^+|H_2$ electrodes. The two aqueous solutions of HCl connected by a liquid junction are of different concentrations (see Fig. 4.4). The changes taking place when one mole of electrons passes through the outer circuit from left to right will be analysed as for the cell discussed in Section 4.1.1.

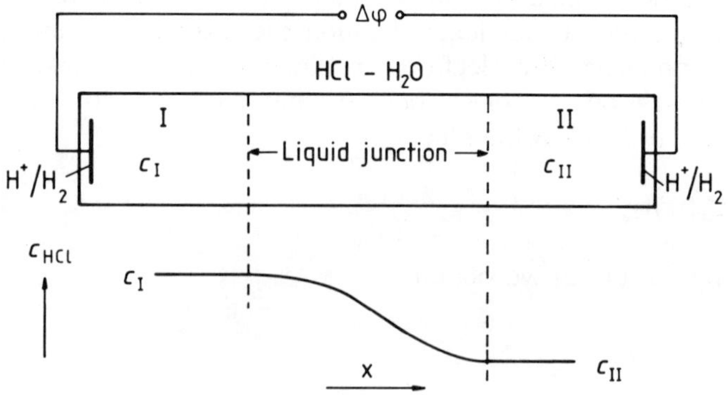

Figure 4.4. An HCl–H_2O concentration cell with $H^+|H_2$ electrodes.

At the electrodes the chemical reactions are the following:

left-hand-side electrode: $\frac{1}{2}\, H_2(g) \rightarrow H^+ + e^-$
right-hand-side electrode: $H^+ + e^- \rightarrow \frac{1}{2}\, H_2(g)$

One half mole H_2 is transferred from the left-hand-side electrode to the righthand-side one, while one mole H^+ is removed from the electrolyte at the electrode on the right-hand side and added to the electrolyte at the electrode on the left-hand side.

The aqueous solution stays electrically neutral and the transfer of one mole H^+ from right to left is compensated for by the migration of Cl^- to the left and of H^+ to the right in the electrolyte, Cl^- and H^+ being the current carriers in the electrolyte. The sum of the transference numbers is equal to one, $t_{H^+} + t_{Cl^-} = 1$. The total changes caused by the transfer of charge in electrolyte I on the lefthand side may be written as follows:

$$\underbrace{(1 - t_{H^+,I})}_{t_{Cl^-,I}} \text{ mol } H^+ + t_{Cl^-,I} \text{ mol } Cl^- = + t_{Cl^-,I} \text{ mol HCl} \tag{4.15}$$

and the total changes in electrolyte II on the left-hand side is correspondingly

$$\underbrace{(-1 + t_{H^+,II})}_{t_{Cl^-,II}} \text{ mol } H^+ - t_{Cl^-,II} \text{ mol } Cl^- = - t_{Cl^-,II} \text{ mol HCl} \tag{4.16}$$

The flux equations for this cell may be written as follows:

$$J'_{HCl} = -L'_{11} \nabla \mu_{HCl} - L'_{12} \nabla \varphi' \tag{4.17a}$$

$$j' = -L'_{21} \nabla \mu_{HCl} - L'_{22} \nabla \varphi' \tag{4.17b}$$

The transfer of $H_2(g)$ between the electrodes does not enter the flux equations as it takes place at constant chemical potential.

The phenomenological coefficients will be compared for two cells that differ only in electrodes, their electrolytes having identical compositions. The electric conductivity of the electrolyte is independent of the electrodes and thus:

$$L'_{22} = L_{22} \tag{4.18}$$

The Hittorf method may again be used to determine the transference numbers. With $H^+|H_2$ electrodes, the net transport of HCl will be from right to left, the opposite direction of the net transport of HCl when using AgCl|Ag electrodes. Thus:

$$J'_{HCl}/j' = L'_{12}/L_{22} = t_{HCl} \tag{4.19}$$

With $H^+|H_2$ electrodes, the transference coefficient is negative and equal to minus the transference number of Cl^-, $t_{HCl} = -\ t_{Cl^-}$ (cf. eqs. (4.15) and (4.16)).

To determine L'_{11} the pure diffusion, $J'_{HCl} = -\ l'_{11}\nabla\mu_{HCl}$, can be studied. Pure diffusion is electrode independent and thus:

$$l'_{11} = l_{11} \tag{4.20}$$

By analogy with eq. (4.10):

$$l'_{11} = L'_{11} - L'_{12}L'_{21}/L_{22} \tag{4.21}$$

and L'_{11} can be found from the known values of l'_{11} and $L'_{12} = L'_{21}$.

As discussed in Section 4.1.1, the short-circuited cell with zero electric potential can be studied, and it is found that

$$(J'_{HCl}/j')_{\nabla\varphi' = 0} = -\ 1$$

which gives

$$L'_{11} = -\ L'_{21} \tag{4.22}$$

Interpretation of coefficients in terms of mobilities. The current carriers in the electrolyte are the ions, and the conductivity of the electrolyte, κ, is determined by the number of current carriers in the volume and by their mobilities:

$$\kappa = c_{H^+}F\,u_{H^+} + c_{Cl^-}F\,u_{Cl^-}$$

The concentration, $c_{H^+} = c_{Cl^-} = c$, is given as the number of moles per m^3, and F is the Faraday constant, 96,500 A s mol^{-1}. The *mobility, u,* is defined as *the velocity of the ion divided by the field strength,* and its dimension is m s^{-1}/(V m^{-1}) = $m^2\ s^{-1}\ V^{-1}$. Thus the dimension for κ is A V^{-1} m^{-1}. By eq. (4.6) we have $L_{22} = \kappa/F^2$, and thus:

$$L_{22} = \frac{c}{F}(u_{H^+} + u_{Cl^-}) \tag{4.23}$$

The transference numbers can also be expressed by the mobilities of the ions:

$$\frac{L_{12}}{L_{22}} = t_{H^+} = \frac{c_{H^+}u_{H^+}}{c_{H^+}u_{H^+} + c_{Cl^-}u_{Cl^-}} = \frac{u_{H^+}}{u_{H^+} + u_{Cl^-}} \tag{4.24a}$$

$$-\frac{L'_{12}}{L_{22}} = t_{Cl^-} = \frac{c_{Cl^-}u_{Cl^-}}{c_{H^+}u_{H^+} + c_{Cl^-}u_{Cl^-}} = \frac{u_{Cl^-}}{u_{H^+} + u_{Cl^-}} \tag{4.24b}$$

Combination with eq. (4.23) gives

$$L_{12} = \frac{c}{F}u_{H^+} \quad \text{and} \quad -L'_{12} = \frac{c}{F}u_{Cl^-} \tag{4.25}$$

By eqs. (4.14) and (4.22) we have $L_{11} = L_{12}$ and $L'_{11} = -L'_{12}$. The coefficient l_{11}, given by eq. (4.10) or eq. (4.21), can thus be expressed by

$$l_{11} = L_{22}t_{H^+}t_{Cl^-} = \frac{c}{F}\frac{u_{H^+}u_{Cl^-}}{u_{H^+} + u_{Cl^-}} \tag{4.26}$$

The relation between the chemical potential and the concentration. When phenomenological coefficients are expressed by concentrations, it is convenient also to express forces by concentrations. The force $X_{HCl} = -\nabla\mu_{HCl}$ is related to the *activity* of HCl, a_{HCl}, by the equation defining the activity, $RT\ln a_{HCl} = \mu_{HCl} - \mu^{o}_{HCl}$, where μ^{o}_{HCl} refers to the standard state of HCl. The activity of a strong electrolyte, such as HCl, in a pure or mixed solution, is related to the concentration of the ions H^+ and Cl^- by the *activity coefficient*, y_{HCl}, $a_{HCl} = c_{H^+}c_{Cl^-}y_{HCl}$. In a pure solution $c_{H^+} = c_{Cl^-} = c$ and $a_{HCl} = c^2y_{HCl}$. The activity coefficient approaches unity when c approaches zero. (Frequently the mean activity coefficient $y_\pm = \sqrt{y_{HCl}}$ is used). Over a concentration range where y_{HCl} may be considered to be approximately constant:

$$X_{HCl} = -\,d\mu_{HCl}/dx = -\,RT\frac{d}{dx}\ln(c^2 y_{HCl}) = -\,\frac{2RT}{c}\frac{dc}{dx} \quad (4.27a)$$

or

$$X_{HCl} = -\,\nabla\mu_{HCl} = -\,\frac{2RT}{c}\nabla c$$

when gradients occur in more than one direction.

When the solute is an undissociated molecule, A, the activity of A is related to its concentration by the equation $a_A = cy_A$. Over a concentration range where y_A may be considered as approximately constant, the following force is obtained:

$$X_A = -\,d\mu_A/dx = -\,RT\frac{d}{dx}\ln(cy) = -\,\frac{RT}{c}\frac{dc}{dx} \quad (4.27b)$$

or

$$X_A = -\,\nabla\mu_A = -\,\frac{RT}{c}\nabla c$$

The diffusion coefficient of Fick's laws of diffusion. Fick's first law of diffusion, formulated for the diffusion of HCl in the electrolyte, is

$$J_{HCl} = -\,D\,dc_{HCl}/dx \quad (4.28)$$

where D is *Fick's diffusion coefficient.* A flux equation for pure diffusion of HCl may be written as (cf. eq. (4.9))

$$J_{HCl} = -\,l_{11}\,d\mu_{HCl}/dx \quad (4.29)$$

Using eqs. (4.26) and (4.27a) we obtain from eq. (4.29)

$$J_{HCl} = -\,\frac{2RT}{F}\,\frac{u_{H^+}u_{Cl^-}}{u_{H^+} + u_{Cl^-}}\,dc_{HCl}/dx \quad (4.30)$$

A comparison of eqs. (4.28) and (4.30) gives

$$D = \frac{2RT}{F}\frac{u_{H^+}u_{Cl^-}}{u_{H^+}+u_{Cl^-}} \tag{4.31}$$

Nernst–Planck flux equations.[3, 4] These equations operate with fluxes of single ions and with gradients in electrochemical potential. The Nernst–Planck equations can be written as follows:

$$J_{H^+} = -\frac{c_{H^+}u_{H^+}}{F}\nabla\tilde{\mu}_{H^+} = -\frac{c_{H^+}u_{H^+}}{F}(\nabla\mu_{H^+} + \nabla\psi) \tag{4.32a}$$

$$J_{Cl^-} = -\frac{c_{Cl^-}u_{Cl^-}}{F}\nabla\tilde{\mu}_{Cl^-} = -\frac{c_{Cl^-}u_{Cl^-}}{F}(\nabla\mu_{Cl^-} - \nabla\psi) \tag{4.32b}$$

By analysis of the changes taking place in the cell (eqs. (4.2) and (4.3)) we can see that J_{H^+} equals J_{HCl} of flux eq. (4.4a), while J_{Cl^-} represents the migration of Cl^- ions in the solution. From eq. (4.25) we see that $c_{H^+}u_{H^+}/F = L_{12}$ while $c_{Cl^-}u_{Cl^-}/F = -L'_{12}$.

The gradient in electrochemical potential for an ion is defined by the following equation:

$$\nabla\tilde{\mu}_i = \nabla\mu_i + z_i\nabla\psi \tag{4.33}$$

where z_i is the charge of the ion and $\nabla\psi$ is the gradient in electrostatic potential in the liquid phase. The single ion chemical potential gradients are defined by

$$\nabla\mu_{H^+} + \nabla\mu_{Cl^-} = \nabla\mu_{HCl} \tag{4.34}$$

The quantities $\nabla\mu_{H^+}$, $\nabla\mu_{Cl^-}$ and $\nabla\psi$ are all unmeasurable.

When using AgCl|Ag electrodes, the forces $\nabla\tilde{\mu}_{H^+}$ and $\nabla\tilde{\mu}_{Cl^-}$ can be expressed by $\nabla\mu_{HCl}$ and $\nabla\varphi$. The dissipation function is independent of the choice of forces, and thus

$$T\Theta = -J_{H^+}\nabla\tilde{\mu}_{H^+} - J_{Cl^-}\nabla\tilde{\mu}_{Cl^-} = -J_{HCl}\nabla\mu_{HCl} - j\nabla\varphi \tag{4.35}$$

The relations between fluxes are $J_{H^+} - J_{Cl^-} = j$ and $J_{H^+} = J_{HCl}$. Between the forces we have the relations $\nabla\tilde{\mu}_{H^+} + \nabla\tilde{\mu}_{Cl^-} = \nabla\mu_{H^+} + \nabla\mu_{Cl^-} = \nabla\mu_{HCl}$.

Thus

$$T\Theta = -J_{\mathrm{H}^+}(\nabla\tilde{\mu}_{\mathrm{H}^+} + \nabla\tilde{\mu}_{\mathrm{Cl}^-}) + j\nabla\tilde{\mu}_{\mathrm{Cl}^-}$$
$$= -J_{\mathrm{HCl}}\nabla\mu_{\mathrm{HCl}} - j\nabla\varphi$$

By inspection of the above equation it is found that

$$\nabla\tilde{\mu}_{\mathrm{Cl}^-} = -\nabla\varphi \tag{4.36}$$

and

$$\nabla\tilde{\mu}_{\mathrm{H}^+} = \nabla\mu_{\mathrm{HCl}} + \nabla\varphi \tag{4.37}$$

A similar reasoning can be carried out for a cell using $H^+|H_2$ electrodes, to obtain

$$\nabla\tilde{\mu}_{\mathrm{Cl}^-} = \nabla\mu_{\mathrm{HCl}} - \nabla\varphi' \tag{4.38}$$

and

$$\nabla\tilde{\mu}_{\mathrm{H}^+} = +\nabla\varphi' \tag{4.39}$$

Thus the electrochemical potentials are expressed by the measurable quantities $\nabla\mu_{\mathrm{HCl}}$ and $\nabla\varphi$ or $\nabla\varphi'$. Notice that

$$\nabla\mu_{\mathrm{HCl}} = \nabla\varphi' - \nabla\varphi \tag{4.40}$$

For only one salt in solution the Nernst–Planck equation is valid, with the reservation that diffusion in form of ion pairs be negligibly small.

4.1.3 *A concentration cell containing the chloride of a divalent cation, AgCl|Ag electrodes*

Let us consider the following cell at constant temperature and pressure:

$$\mathrm{Ag(s)|AgCl(s)|CaCl_2(aq,c_I)||(CaCl_2(aq,c_{II})|AgCl(s)|Ag(s)}$$

The changes taking place when one mole of electrons passes from left to right in the outer circuit may be studied in the same way as for the previous cells.

Total changes on the left-hand side:

$$(-1 + t_{Cl^-,I}) \text{ mol } Cl^- - \tfrac{1}{2}\, t_{Ca^{2+},I} \text{ mol } Ca^{2+} = -\tfrac{1}{2}\, t_{Ca^{2+},I} \text{ mol } CaCl_2 \tag{4.41}$$

Total changes on the right-hand side:

$$(1 - t_{Cl^-,II}) \text{ mol } Cl^- + \tfrac{1}{2}\, t_{Ca^{2+},II} \text{ mol } Ca^{2+} = +\tfrac{1}{2}\, t_{Ca^{2+},II} \text{ mol } CaCl_2 \tag{4.42}$$

(Cf. eqs. (4.2) and (4.3).)

The transference number for Ca^{2+} ions is defined as the fraction of electric charges transferred by means of the Ca^{2+} ions, i.e. it refers to the number of *equivalents* Ca^{2+} transferred. Since transference numbers refer to equivalents, the sum of the transference numbers is equal to unity:

$$\Sigma t_+ + \Sigma t_- = 1 \tag{4.43}$$

The transference coefficients, however, refer to the number of moles of neutral salts (see eq. (4.7)). In this cell we have

$$t_{CaCl_2} = \tfrac{1}{2}\, t_{Ca^{2+}} \tag{4.44}$$

The flux equations expressing the changes are

$$J_{CaCl_2} = -L_{11}\nabla\mu_{CaCl_2} - L_{12}\nabla\varphi \tag{4.45a}$$

$$j = -L_{21}\nabla\mu_{CaCl_2} - L_{22}\nabla\varphi \tag{4.45b}$$

As for the previous cells, the experiments are arranged such that gradients occur only in the direction of the x-axis, when studying the phenomenological coefficients.

Again L_{22} is found from conductivity measurements when $d\mu_{CaCl_2}/dx = 0$. The coefficient L_{22} can be interpreted in terms of mobilities of the ions, $u_{Ca^{2+}}$ and u_{Cl^-}, and the concentration of $CaCl_2$, c:

$$L_{22} = \frac{2c}{F}(u_{Ca^{2+}} + u_{Cl^-}) \tag{4.46}$$

Further, the Hittorf method is used to find t_{CaCl_2}, and from eqs. (4.45):

$$J_{CaCl_2}/j = L_{12}/L_{22} = t_{CaCl_2} = \tfrac{1}{2}\, t_{Ca^{2+}} = \tfrac{1}{2}\, \frac{u_{Ca^{2+}}}{u_{Ca^{2+}} + u_{Cl^-}} \tag{4.47}$$

When short-circuiting the cell, such that $d\varphi/dx = 0$, we obtain the following from eqs. (4.45):

$$J_{CaCl_2}/j = \tfrac{1}{2} = L_{11}/L_{21} \tag{4.48}$$

For pure diffusion we have,

$$J_{CaCl_2} = -\, l_{11} \nabla \mu_{CaCl_2} \tag{4.49}$$

where $l_{11} = L_{11} - L_{12}L_{21}/L_{22}$ (cf. eq. (4.10)).

Combining eq. (4.10) with eqs. (4.47) and (4.48), and bearing in mind that $t_{Cl^-} = 1 - t_{Ca^{2+}}$ and $L_{12} = L_{21}$, we obtain

$$l_{11} = \tfrac{1}{4}\, L_{22} t_{Ca^{2+}} t_{Cl^-} \tag{4.50}$$

The coefficient L_{12} can also be expressed by concentration and mobility:

$$L_{12} = \frac{c}{F} u_{Ca^{2+}} \tag{4.51}$$

4.1.4 *A concentration cell containing two different chlorides, AgCl|Ag electrodes*

Let us consider the following cell at constant temperature and pressure:

$$\mathrm{Ag(s)|AgCl(s)|HCl(aq,}c_{\mathrm{HCl,I}}\mathrm{),NaCl(aq,}c_{\mathrm{NaCl,I}}\mathrm{)||}$$
$$\mathrm{HCl(aq,}c_{\mathrm{HCl,II}}\mathrm{),NaCl(aq,}c_{\mathrm{NaCl,II}}\mathrm{)|AgCl(s)|Ag(s)}$$

The changes taking place when one mole of electrons passes from left to right in the outer circuit may be studied in the same way as for the previous cells.

Total changes on the left-hand side:

$$(-1 + t_{Cl^-,I})\text{ mol } Cl^- - t_{H^+,I}\text{mol } H^+ - t_{Na^+,I}\text{ mol } Na^+ = -t_{H^+,I}\text{ mol HCl} - t_{Na^+,I}\text{ mol NaCl} \quad (4.52)$$

Total changes on the right-hand side:

$$(1 - t_{Cl^-,II})\text{ mol } Cl^- + t_{H^+,II}\text{ mol } H^+ + t_{Na^+,II}\text{ mol } Na^+ =$$

$$t_{H^+,II}\text{ mol HCl} + t_{Na^+,II}\text{ mol NaCl} \quad (4.53)$$

The flux equations expressing these changes are the following:

$$J_{HCl} = -L_{11}\nabla\mu_{HCl} - L_{12}\nabla\mu_{NaCl} - L_{13}\nabla\varphi \quad (4.54a)$$

$$J_{NaCl} = -L_{21}\nabla\mu_{HCl} - L_{22}\nabla\mu_{NaCl} - L_{23}\nabla\varphi \quad (4.54b)$$

$$j = -L_{31}\nabla\mu_{HCl} - L_{32}\nabla\mu_{NaCl} - L_{33}\nabla\varphi \quad (4.54c)$$

Again the experiments are arranged so that gradients occur only in the direction of the x-axis, when studying the phenomenological coefficients. The same type of experiments are used as for the previous cells.

The coefficient L_{33} is found from conductivity measurements when $d\mu_{HCl}/dx = 0$ and $d\mu_{NaCl}/dx = 0$. Using a Hittorf method we find that

$$J_{HCl}/j = L_{13}/L_{33} = t_{HCl} = t_{H^+} \quad (4.55)$$

$$J_{NaCl}/j = L_{23}/L_{33} = t_{NaCl} = t_{Na^+} \quad (4.56)$$

As for the simpler case with only one salt in solution, the system may be short-circuited in such a way that $d\varphi/dx = 0$ (cf. Fig. 4.3). When $d\varphi/dx = 0$ and the migration of neutral HCl and NaCl pairs is negligible, transfer of Cl^- can only take place by the electrode reactions. Then we have

$$J_{HCl} + J_{NaCl} \approx j \quad (4.57)$$

When $d\varphi/dx = 0$ and, at the same time, $d\mu_{NaCl}/dx = 0$, we obtain from eqs. (4.54) and (4.57):

$$L_{11} + L_{21} = L_{31} \quad (4.58)$$

and similarly when $d\varphi/dx = 0$ and $d\mu_{HCl}/dx = 0$, we obtain

$$L_{12} + L_{22} = L_{32} \tag{4.59}$$

Since coefficients are independent of forces, eqs. (4.58) and (4.59) are also valid when the forces are not zero.

Nernst–Planck flux equations. If we assume the cross coefficients $L_{12} = L_{21}$ to be very small, eq. (4.58) reduces to $L_{11} \approx L_{31}$ and eq. (4.59) to $L_{22} \approx L_{32}$. Then eqs. (4.54a) and (4.54b) reduce to

$$J_{HCl} = -L_{13}(\nabla\mu_{HCl} + \nabla\varphi) \tag{4.60a}$$

$$J_{NaCl} = -L_{23}(\nabla\mu_{NaCl} + \nabla\varphi) \tag{4.60b}$$

These equations are equivalent to the Nernst–Planck equations for the system since $J_{H^+} = J_{HCl}$ and $J_{Na^+} = J_{NaCl}$ (cf. eqs. (4.32) and (4.35)).

Coefficients, transference numbers and transference coefficients may be expressed by means of concentrations and mobilities. The concentration in the electrolyte is $c = c_{HCl} + c_{NaCl}$ and the mole fractions are $x_{H^+} = c_{HCl}/c$, $x_{Na^+} = c_{NaCl}/c$ and $x_{Cl^-} = 1$. The expressions for the coefficients are

$$L_{33} = \frac{c}{F}(x_{H^+}u_{H^+} + x_{Na^+}u_{Na^+} + u_{Cl^-}) = \frac{c}{F}\Sigma_i x_i u_i \tag{4.61}$$

The summation is over all ionic species. Furthermore:

$$L_{13} = \frac{c}{F}x_{H^+}u_{H^+} \quad \text{and} \quad L_{23} = \frac{c}{F}x_{Na^+}u_{Na^+} \tag{4.62}$$

The transference coefficients for HCl and NaCl are equal to the transference numbers for H^+ and Na^+ respectively:

$$t_{HCl} = t_{H^+} = L_{13}/L_{33} = \frac{x_{H^+}u_{H^+}}{\Sigma_i x_i u_i} \tag{4.63}$$

$$t_{NaCl} = t_{Na^+} = L_{23}/L_{33} = \frac{x_{Na^+}u_{Na^+}}{\Sigma_i x_i u_i} \tag{4.64}$$

Since $t_{H^+} + t_{Na^+} + t_{Cl^-} = 1$:

$$t_{Cl^-} = \frac{u_{Cl^-}}{\Sigma_i x_i u_i} \tag{4.65}$$

Pure diffusion. With two chlorides in the solution, the diffusion process is more complex than with only one chloride in the solution. The flux of HCl may be viewed as composed of two parts, the interdiffusion of H^+ and Na^+ and the diffusion of the neutral compound HCl and similarly for the flux of NaCl.

By eliminating $\nabla\varphi$ in eqs. (4.54), we obtain in replacement of (4.54a):

$$J_{HCl} = -(L_{11} - L_{13}L_{31}/L_{33})\nabla\mu_{HCl} - (L_{12} - L_{13}L_{32}/L_{33})\nabla\mu_{NaCl} + (L_{13}/L_{33})j \tag{4.66}$$

and similarly for J_{NaCl} in replacement of eq. (4.54b).

For pure diffusion $j = 0$. Furthermore, the cross coefficient $L_{12} = L_{21}$ may be considered to be negligibly small, and thus:

$$J_{HCl} = -(L_{13} - L_{13}L_{31}/L_{33})\nabla\mu_{HCl} + (L_{13}L_{32}/L_{33})\nabla\mu_{NaCl}$$

By introducing the transference numbers, $t_{H^+} = L_{13}/L_{33}$, $t_{Na^+} = L_{23}/L_{33}$ and $t_{Cl^-} = 1 - t_{H^+} - t_{Na^+}$, we obtain

$$J_{HCl} = -L_{33}t_{H^+}t_{Na^+}(\nabla\mu_{HCl} - \nabla\mu_{NaCl}) - L_{33}t_{H^+}t_{Cl^-}\nabla\mu_{HCl} \tag{4.67}$$

The first term of the equation represents the interdiffusion of H^+ and Na^+, while the second term represents the diffusion of HCl. A similar equation is obtained for J_{NaCl}.

The phenomenological coefficients of the flux equations, how they can be determined experimentally and how they are interrelated under given conditions have been considered in this section. The emf of concentration cells will be treated in the next section.

4.2 *The emf of Concentration Cells*

In electrochemistry one commonly needs to calculate the emf of electrochemical cells, i.e. $\Delta\varphi$ for $j = 0$. Two methods will be discussed. By the first method one integrates the equation for $\nabla\varphi$. By the second method one calculates the changes in composition upon charge transfer over all regions of the cell and the corresponding ΔG which can be related to $\Delta\varphi$.

4.2.1 *Integration of the equation for* $\nabla\varphi$

Let us consider the concentration cell of Section 4.1.1:

$$Ag(s)|AgCl(s)|HCl(aq,c_I)||HCl(aq,c_{II})|AgCl(s)|Ag(s)$$

With gradients in the x-direction only, we have for $j = 0$ (see eq. (4.8))

$$\Delta\varphi = \int_{(I)}^{(II)} \nabla\varphi dx = -\int t_{HCl} d\mu_{HCl} = -\int t_{H^+} d\mu_{HCl} \tag{4.68}$$

In order to carry out the integration, one must know the corresponding values of t_{H^+} and μ_{HCl} over the concentration interval $(c_I - c_{II})$. The t_{H^+} is a function of μ_{HCl} because mobilities change slightly with concentration. Usually this function is not known, and the integration must be carried out numerically from experimental values.

If the AgCl|Ag electrodes are replaced by $H^+|H_2$ electrodes, (cf. the cell of Section 4.1.2) we obtain

$$\Delta\varphi' = \int_{(I)}^{(II)} \nabla\varphi' dx = \int t_{Cl^-} d\mu_{HCl} \tag{4.69}$$

The difference between the emfs of the two cells that have identical electrolytes but different electrodes, $H^+|H_2$ and AgCl|Ag respectively, is

$$\Delta\varphi' - \Delta\varphi = \int (t_{Cl^-} + t_{H^+})\, d\mu_{HCl} = \Delta\mu_{HCl} \tag{4.70}$$

(Cf. eq. (4.40)).

In a similar way the emf, $\Delta\varphi$, for a cell containing two electrolytes, HCl and NaCl, can be found (cf. the cell of Section

4.1.4). When $j = 0$, eqs. (4.54c), (4.55) and (4.56) give $\nabla\varphi = -t_{H^+}\nabla\mu_{HCl} - t_{Na^+}\nabla\mu_{NaCl}$. With gradients in the x-direction only:

$$\Delta\varphi = -\int_{(I)}^{(II)} (t_{H^+}d\mu_{HCl} + t_{Na^+}d\mu_{NaCl}) \tag{4.71}$$

The t_{H^+} and t_{Na^+} are functions of the mole fractions of H^+ and Na^+ respectively, as well as functions of mobilities which change slightly with concentration (see eqs. (4.63) and (4.64)). There is an infinite number of concentration paths through the liquid junction, from $c_{HCl,I}$ and $c_{NaCl,I}$ to $c_{HCl,II}$ and $c_{NaCl,II}$. In order to integrate eq. (4.71) one must know the concentration profile i.e. how the concentrations vary with distance.

4.2.2 *Changes in Gibbs energy by diffusion and charge transfer. Relation to the emf of the cell*

By this alternative method of calculating the emf, the local changes in composition upon charge transfer are studied in more detail.[5] A similar method was used by Hertz.[6]

For a *reversible* cell the emf is derived from equilibrium thermodynamics, and we have

$$\Delta G + \Delta\varphi = 0 \tag{4.72}$$

where ΔG is the change in Gibbs energy of the cell per faraday transferred. Some cells without a liquid junction may behave reversibly under certain conditions, e.g. a cell where the two electrolytes are separated by an ion exchange membrane, or a cell with only one electrolyte, but different electrodes.

In a cell with a liquid junction, an *irreversible* diffusion of matter takes place. Both charge transfer and diffusion contribute to the change in Gibbs energy. We shall see that these two contributions are separable.

Let us again consider the concentration cell of Section 4.1.1:

$$Ag(s)|AgCl(s)|HCl(aq,c_I)||HCl(aq,c_{II})|AgCl(s)|Ag(s)$$

In Section 4.1.1 it was shown that the flux of matter through the liquid junction is composed of a diffusion term and a charge-transfer term; as expressed in eq. (4.9): $J_{HCl} = -l_{11}\nabla\mu_{HCl} + t_{HCl}j$. The diffusion takes place continuously, even without charge transfer ($j = 0$).

The change in Gibbs energy in the *liquid junction* depends both on time and on the charge transferred, whereas the changes at the electrodes depend only on charge transferred when the electrodes are reversible.

A closer look will be taken at the changes taking place in the liquid junction and, for the sake of simplicity, the experiment will be arranged with gradients in the x-direction only (see Fig. 4.5).

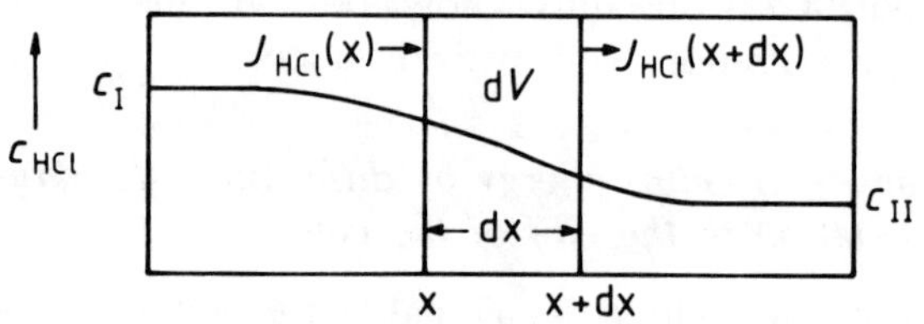

Figure 4.5. Flux into and out of a volume element in the liquid junction when the concentration gradient varies.

Consider a narrow volume element, dV, between planes through x and x + dx (see Fig. 4.5). The change in concentration of HCl in this volume element per unit time is equal to the difference between the flux of HCl into and out of the volume element, divided by the thickness of the volume element, dx:

$$dc_{HCl}/dt = \frac{J_{HCl}(x) - J_{HCl}(x+dx)}{dx} = -\,dJ_{HCl}/dx \tag{4.73}$$

The corresponding change in the Gibbs energy of the volume element in the liquid junction, $d(dG_{l.j.})$, is

$$d(dG_{l.j.}) = dV\,dc_{HCl}\mu_{HCl} = A dx\, dc_{HCl}\, \mu_{HCl} \tag{4.74}$$

where A is the cross-section of the volume element and Adx is equal to the volume dV.

The change in concentration of HCl, dc_{HCl},can be expressed by the derivative of eq. (4.9):

$$dc_{HCl} = -(dJ_{HCl}/dx)dt = -\frac{d}{dx}(-l_{11}d\mu_{HCl}/dx + t_{HCl}j)dt$$

This expression for dc_{HCl} may be introduced into eq. (4.74), giving

$$d(dG_{l.j.}) = A\mu_{HCl}d(l_{11}d\mu_{HCl}/dx)dt - A\mu_{HCl}dt_{HCl}jdt$$

and since the charge transferred is $dQ = Ajdt$:

$$d(dG_{l.j.}) = A\mu_{HCl}d(l_{11}d\mu_{HCl}/dx)dt - \mu_{HCl}dt_{HCl}dQ \tag{4.75}$$

The change in Gibbs energy over the whole liquid junction, $dG_{l.j.}$, is obtained by integration:

$$dG_{l.j.} = \left[\int_{(I)}^{(II)} A\mu_{HCl}d(l_{11}d\mu_{HCl}/dx)\right]dt - \left[\int_{(I)}^{(II)} \mu_{HCl}dt_{HCl}\right]dQ \tag{4.76}$$

In addition to $dG_{l.j.}$ there are changes in Gibbs energy at the two electrodes, dG_{el}:

$$dG_{el} = (-\mu_{HCl,I}t_{HCl,I} + \mu_{HCl,II}t_{HCl,II})dQ \tag{4.77}$$

The change in Gibbs energy over the whole cell is $dG = dG_{l.j.} + dG_{el}$, the sum of eqs. (4.76) and (4.77). Inspection of the equations shows that dG may be written as the sum of a time-dependent term $dG(t)$ and a charge-dependent term $dG(Q)$:

$$dG = dG(t) + dG(Q) \tag{4.78}$$

The first term expresses the change in Gibbs energy caused by diffusion. It depends on time, but is independent of charge transfer:

$$dG(t) = \left[\int_{(I)}^{(II)} A\mu_{HCl}d(l_{11}d\mu_{HCl}/dx)\right]dt \tag{4.79}$$

The second term expresses the change in Gibbs energy caused by charge transfer. It depends on charge, but is independent of time:

$$dG(Q) = -\left[\int_{(I)}^{(II)} \mu_{HCl} dt_{HCl}\right] dQ + \left[\mu_{HCl,II} t_{HCl,II.} - \mu_{HCl,I} t_{HCl,I}\right] dQ \tag{4.80}$$

Thus the changes with time and the changes with charge transfer are separable.

The first term on the righthand side of eq. (4.80) represents the change in Gibbs energy in the liquid junction and it is independent of the choice of electrodes. The second term represents the change in Gibbs energy at the electrodes.

Let us consider a closed cell with compensation of the electric potential by an external voltage source. By very small changes in the emf of the external voltage source, a charge dQ may be transported either way through the cell, and the electric work, $-\Delta\varphi dQ$, may be converted to Gibbs energy, $dG(Q)$. Since j is very small in this type of measurement, the Joule heat, Rj^2, is negligible compared to $dG(Q)/dt$, which is proportional to the first power of j (cf. discussion in Appendix 2 (Fig. A2.1)). We obtain (cf. eq. A2.32)

$$\Delta\varphi dQ + dG(Q) = 0 \tag{4.81}$$

The change in Gibbs energy per faraday transferred is $\Delta G(Q) = dG/dQ = -\Delta\varphi$, or

$$\Delta G(Q) + \Delta\varphi = 0 \tag{4.82}$$

Compare eq. (4.82) with the corresponding equation for a reversible cell (eq. (4.72)). Equation (4.82) is valid for irreversible and reversible cells. When there is no irreversible diffusion in a cell, the term $dG(t)$ of eq. (4.78) is equal to zero. The cell is reversible and $\Delta G = \Delta G(Q)$.

From eq. (4.80):

$$\Delta G(Q) = dG(Q)/dQ = \Big|_{(I)}^{(II)} \mu_{HCl} t_{HCl} - \int_{(I)}^{(II)} \mu_{HCl}\, dt_{HCl} = -\Delta\varphi \tag{4.83}$$

By the rules of integration by parts $\Delta\varphi = -\int_{(I)}^{(II)} t_{HCl}\mathrm{d}\mu_{HCl}$, the same expression as eq. (4.68).

For a cell containing several salts in solution, each salt present will contribute to the change in Gibbs energy with a time-dependent term and a charge-dependent term. Let us consider the cell discussed in Section 4.1.4:

$$\mathrm{Ag(s)|AgCl(s)|HCl(aq,}c_{\mathrm{HCl,I}}\mathrm{),\ NaCl(aq,}c_{\mathrm{NaCl,I}}\mathrm{)||}$$
$$\mathrm{HCl(aq,}c_{\mathrm{HCl,II}}\mathrm{),\ NaCl(aq,}c_{\mathrm{NaCl,II}}\mathrm{)|AgCl(s)|Ag(s)}$$

The elimination of $\nabla\varphi$ in eqs. (4.54) gives

$$J_{HCl} = -l_{11}\nabla\mu_{HCl} - l_{12}\nabla\mu_{NaCl} + t_{HCl}j \tag{4.84a}$$

$$J_{NaCl} = -l_{21}\nabla\mu_{HCl} - l_{22}\nabla\mu_{NaCl} + t_{NaCl}j \tag{4.84b}$$

where $l_{ij} = L_{ij} - L_{i3}L_{j3}/L_{33}$ (cf. eq. (4.66)).

Using a treatment similar to that for the simpler case with only one salt in solution, the following expressions are obtained for $\mathrm{d}G(t)$ and $\mathrm{d}G(Q)$:

$$\mathrm{d}G(t) = \left[\int_{(I)}^{(II)} A\mu_{HCl}\mathrm{d}\{l_{11}\mathrm{d}\mu_{HCl}/\mathrm{d}x + l_{12}\mathrm{d}\mu_{NaCl}/\mathrm{d}x\} + \int_{(I)}^{(II)} A\mu_{NaCl}\mathrm{d}\{l_{22}\mathrm{d}\mu_{NaCl}/\mathrm{d}x + l_{21}\mathrm{d}\mu_{HCl}/\mathrm{d}x\}\right]\mathrm{d}t \tag{4.85}$$

$$\mathrm{d}G(Q) = \left[\Big|_{(I)}^{(II)} (\mu_{HCl}t_{HCl} + \mu_{NaCl}t_{NaCl}) - \int_{(I)}^{(II)} (\mu_{HCl}\mathrm{d}t_{HCl} + \mu_{NaCl}\mathrm{d}t_{NaCl})\right]\mathrm{d}Q \tag{4.86}$$

and furthermore

$$\Delta\varphi = -\int_{(I)}^{(II)} (t_{HCl}\mathrm{d}\mu_{HCl} + t_{NaCl}\mathrm{d}\mu_{NaCl}) \tag{4.87}$$

(Cf. eq. (4.71).)

The method derived above for emf calculation can be used for any kind of isothermal cell. Examples of uses are given in Chapter 8.

4.3 *References*

1. Bockris, J. O'M. and Reddy, A. K. N., *Modern Electrochemistry*, Plenum Press, New York 1970.
2. White, J. M., *Physical Chemistry Laboratory Experiments*, Prentice-Hall, New York, 1975.
3. Nernst, W., *Z. Phys. Chem.*, **2**, 613 (1888).
4. Planck, M., *Ann. Phys.*, *40*, 561 (1890).
5. Førland, T., Thulin, L. U. and Østvold, T., *J. Chem. Ed.*, **48**, 741 (1971).
6. Hertz, H. G., *Electrochemistry*, Springer, Berlin, 1980.

4.4 *Exercises*

4.1. *Changes in a concentration cell.* The following cell is given: Ag(s)|AgCl(s)|NaCl(aq,c_I)||NaCl(aq,c_{II})|AgCl(s)|Ag(s). Assume constant transference numbers through the cell. (a) Account for all changes taking place when one mole of electrons passes from left to right in the outer circuit. (b) Write down the flux equations for the cell and give physical interpretations of the phenomenological coefficients.

4.2. *Transference numbers and transference coefficients.* The following cell is given: Zn(Hg)|ZnI_2(aq,c_I)||ZnI_2(aq,c_{II})|Zn(Hg). Assume constant transference numbers through the cell. (a) Account for all changes taking place when one mole of electrons passes from left to right in the outer circuit. (b) What is the difference between transference numbers and transference coefficients? Give the relations between $t_{Zn^{2+}}$, t_{I^-} and t_{ZnI_2}.

4.3. *The emf of concentration cells.* Assume ideal solutions and constant transference numbers for the cells (a) and (b) below. Calculate the emf of each cell at 25°C, when c_I = 0.1 kmol m^{-3} and c_{II} = 0.01 kmol m^{-3}. (a) Ag(s)|AgCl(s)|KCl(aq,c_I)||KCl(aq,c_{II})|AgCl(s)|Ag(s); $t_{K^+} = t_{Cl^-} = \frac{1}{2}$. (b) (Pt)$H_2$(g,1 atm)|HCl(aq,$c_I$)||HCl(aq,$c_{II}$) |$H_2$(g,1 atm)(Pt); $t_{H^+} = 0.83$.

4.4. *Determination of transference number.* The transference number $t_{Zn^{2+}}$ can be determined by comparing the emf for the following two cells:

(1) Zn(s)|$ZnSO_4$(aq,c_I)||$ZnSO_4$(aq,c_{II})|Zn(s),

(2) $Zn(s)|ZnSO_4(aq,c_I)|Hg_2SO_4(s)|Hg(l)|Hg_2SO_4(s)|ZnSO_4(aq,c_{II})|Zn(s)$. Assume constant mobilities for the ions through both cells. (a) Account for all changes taking place in each cell when one mole of electrons passes from left to right in the outer circuit. (b) Derive an expression for the cell potential in each cell. (c) For given values of c_I and c_{II} the emfs of cell (1) and cell (2) were respectively, $E_1 = -0.048$ V and $E_2 = -0.078$ V, measured when $I \approx 0$, at 25°C. Neglect any changes in c_I and c_{II}, and calculate $t_{Zn^{2+}}$ in cell (1).

4.5. *The Nernst–Planck flux equations*. Consider the cell given in Section 4.1.4. (a) The Nernst–Planck flux equation $J_{HCl} = -L_{11}(\nabla\mu_{HCl} + \nabla\varphi)$ is an approximation based on assumptions — which ones? Give a physical interpretation of the assumptions. (b) How many independent coefficients are there in the flux equations when the Nernst–Planck equations are valid in addition to the Onsager reciprocal relations? (c) The Nernst–Planck equation is commonly written $J_{HCl} = -(c_{H^+}u_{H^+}/F)\nabla\tilde{\mu}_{H^+}$. Discuss the relation between the two expressions for the Nernst–Planck equation.

CHAPTER 5

Transport Processes in Ion Exchange Membranes

In Chapter 4 the transport equations for simple concentration cells with liquid junctions were studied. Phenomenological coefficents were interpreted and the emf of the cell was studied. The liquid junction was supposed to be ideal, not adsorbing or hindering the migration of any component selectively.

When the liquid junction is a porous wall with large pore sizes, any adsorption to the solid may be negligible. As the pore sizes diminish, however, the adsorption layer will be of increasing importance, compared to the bulk of solution in the pores. The nature of the solid surface also becomes important as it determines the selectivity of the adsorption. When the pores or channels are of molecular dimensions, the adsorption to the walls may predominate, and the transport of current and matter will be mainly by the migration of adsorbed ions. Such a liquid junction is called a *membrane*. As will be discussed in Section 5.1, there are different types of membranes. It is characteristic for membranes, however, that transport of bulk material is absent or small.

Transport through membranes is of increasing importance in technology and industry. Microfiltration (sterilization), ultrafiltration (pollution control), reverse osmosis and electrodialysis (desalination) and dialysis (artificial kidney) are examples of well-known membrane processes. The substitution of a membrane process for the mercury process in the production of the basic chemicals NaOH and Cl_2 contributes to a reduction of the mercury pollution problem. In analytical chemistry ion exchange membranes have become an important tool. Some of the most important biological processes are based on selective transport through membranes.

Thus many different fields of science and technology may benefit from the understanding and quantitative description of transport processes in membranes.

Many of the transport processes in membranes are complex due to coupling of the transport of different species, to coupling of the transport of charge and matter and to chemical reactions combined with transport.

In this chapter we shall deal with the fundamental equations for transport in membranes. Ion exchange membranes are used as examples to illustrate the theory. Experimental results for a cation exchange membrane are given in Chapter 9.

5.1 *Some Common Types of Membranes*

In the simplest type of membrane large solute particles are prevented from permeating because of size. The pores or channels are sufficiently wide, however, to allow solvent particles (and other small particles) to pass. Such a membrane is called a *semipermeable membrane*. When there is a gradient in the chemical potential of the solvent in the membrane, there will be a flux of solvent particles through the membrane. This flux may give rise to an *osmotic pressure* counteracting the gradient in chemical potential. When applying a high pressure to a solution of large molecules in contact with a semipermeable membrane, one may force a flux of pure solvent through the membrane, leaving the large solute molecules behind. The membrane acts as a *molecular sieve*.

Ion exchange membranes contain *ionogenic groups* and show selectivity with regards to the transport of charged particles (ions). They are often said to be *permselective membranes*. Organic polymers containing negative ionogenic groups combined with exchangeable cations behave as *cation exchangers*. Examples are molecules of the type

$$RSO_3^-Na^+ \quad \text{and} \quad RPO_3^{2-}(H^+)_2$$

where Na^+ and H^+ can be replaced by other cations. Organic polymers containing positive ionogenic groups combined with exchangeable anions behave as *anion exchangers*. Examples are molecules of the type

$$RNH_3{}^+Cl^- \quad \text{and} \quad RN^+(CH_3)_3Br^-$$

where Cl^- and Br^- can be replaced by other anions. Inorganic materials, glasses and ionic crystals may also serve as ion exchangers (ion-selective electrodes).

When the pores and channels of an ion exchange membrane are too narrow to allow transport of bulk solution, the membrane conducts electricity only by the migration of ions via the ionogenic groups. Furthermore, a purely cation-conducting membrane contains only negative ionogenic groups, while a purely anion-conducting membrane contains only positive ionogenic groups. In the membrane the speed of migration may vary widely for different ions with charges of the same sign.

Since an ion exchange membrane has very narrow pores and channels, it acts as a semipermeable membrane in addition to being permselective. Almost all semipermeable membranes, on the other hand, are somewhat permselective, having some preferred adsorptivity to the pore walls.

A *liquid membrane*, as a separate organic phase between two aqueous solutions, may be permselective. For such membranes one obviously cannot talk of pore sizes or localized sites for ionogenic groups. Inorganic ions may dissolve in the organic phase as complexes with the membrane material or other ligands, *ionophores*. Under the influence of an electric field, these complexes may carry the current through the membrane. The selectivity of the membrane depends on the stability constants for the complexes and their mobilities in the membrane.

The structure of *biological membranes* is of particular interest in relation to the study of biological processes. These membranes are formed from *lipids*, large molecules having a polar group at one end. The remainder of the molecule is a long tail consisting of non-polar hydrocarbons. In the membrane the lipids form a double layer. The polar groups are oriented towards the two aqueous phases, outside and inside the cell enclosed by the membrane. The non-polar tails in the interior of the membrane are attracted to one another by weak van der Waals' forces (see Fig. 5.1). The non-polar interior of the membrane acts as a blockage to the transport of the ions. Ions may be transported through charged channels, however, or they may pass through the membrane in form of complexes. The ligands of the complexes,

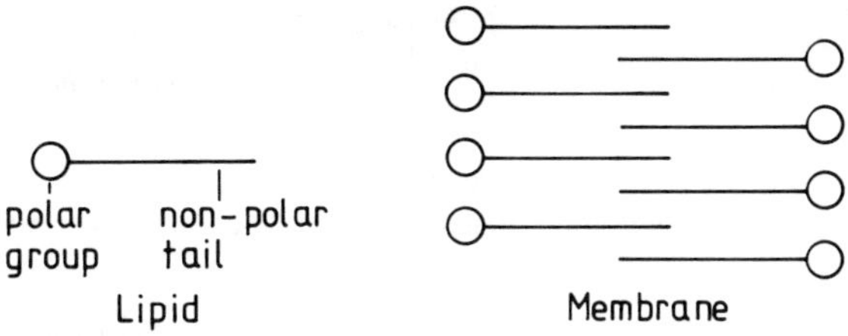

Figure 5.1. Model of a biological membrane.

the *carriers*, increase the solubility of the ions in the non-polar layer.

5.2 *Equilibrium Across the Phase Boundary Membrane – Solution*

A precondition for irreversible thermodynamics is the assumption of local equilibrium (Section 2.2). Thus we must have equilibria across the phase boundaries, between membrane and solution.

We saw in Section 4.1 that the phenomenological coefficients in a liquid junction were expressed by concentrations and mobilities. Similarly the phenomenological coefficients in a membrane can be expressed by concentrations and mobilities. The equilibria across the phase boundaries are of particular interest in this connection, as they link concentrations inside a membrane to the more easily determined concentrations in the adjacent electrolytes.

A schematic illustration of an equilibrium across the phase boundary is given in Fig. 5.2. The electrolyte is a solution of ACl and BCl in H_2O. The membrane is a pure cation exchange membrane and the content of Cl^- ions in the membrane is assumed to be negligible. Since the membrane must be electrically neutral, all negative sites for cations, M^-, must be filled with either A^+ or B^+. The membranes usually have a high content of water in equilibrium with the water in the electrolyte. At equilibrium there is a continuous exchange of cations across the phase boundary, and the equilibrium reaction can be described formally as

$$ACl(aq) + BM \rightleftarrows BCl(aq) + AM \qquad (5.1)$$

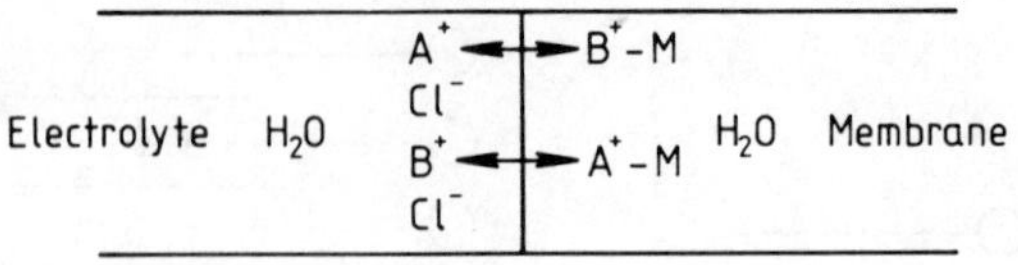

Figure 5.2. Dynamic equilibrium between solution of electrolyte and membrane.

where AM and BM are cation sites in the membrane filled with A^+ and B^+ respectively. Applying the law of chemical equilibrium, we have

$$\frac{a_{BCl}}{a_{ACl}} \cdot \frac{a_{AM}}{a_{BM}} = K \tag{5.2}$$

For dilute electrolyte solutions with known concentrations, activities can be obtained from tables of experimental values,[1, 2] or they can be calculated by means of the Debye–Hückel equation. The content of A^+ and B^+ in the membrane can be determined experimentally, and thus the ratio of the mole fractions, x_{AM}/x_{BM} (where $x_{AM} + x_{BM} = 1$), can be found.

The activity ratio in the membrane may be expressed as

$$\frac{a_{AM}}{a_{BM}} = \frac{f_{AM}}{f_{BM}} \cdot \frac{x_{AM}}{x_{BM}} \tag{5.3}$$

where f_{AM} and f_{BM} are *activity coefficients on a mole fraction basis*. By eqs. (5.2) and (5.3) we obtain

$$\frac{x_{AM}}{x_{BM}} = K \cdot \frac{f_{BM}}{f_{AM}} \cdot \frac{a_{ACl}}{a_{BCl}} \tag{5.4}$$

When experimental values of x_{AM}/x_{BM} are plotted against the ratio a_{ACl}/a_{BCl}, a more or less straight line is obtained. It has been found experimentally that pure cation exchange membranes behave close to ideally for electrolytes containing a mixture of NaCl and KCl. This means that $f_{KM} \approx f_{NaM}$ over the whole range of compositions. For such cases a small number of experimental values will suffice to give accurate values for interpolated composition ratios. The number of experimental values needed for x_{AM}/x_{BM} versus a_{ACl}/a_{BCl} depends on the deviation from ideality and on the accuracy needed for the interpolated values.

5.3 *Concentration Cells with Membranes*

5.3.1 *An HCl–H_2O concentration cell with cation exchange membrane and AgCl|Ag electrodes*

As a simple example we shall consider the following concentration cell at constant temperature and pressure:

$Ag(s)|AgCl(s)|HCl(aq,c_I)|^C|HCl(aq,c_{II})|AgCl(s)|Ag(s)$

The two electrodes are both AgCl|Ag electrodes. The two aqueous solutions of HCl, separated by a cation exchange membrane, $|^C|$ are of different concentration. We assume no concentration gradient in electrolytes I and II (vigorous stirring). The membrane is a cation exchange membrane with a high concentration of HM. A schematic illustration of the cell is given in Fig. 5.3.

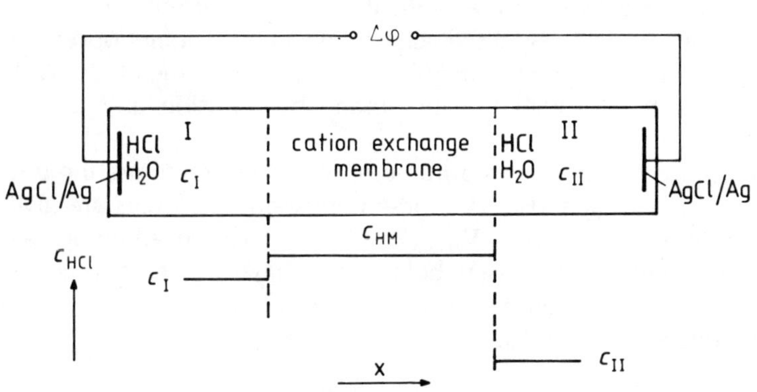

Figure 5.3. An HCl–H_2O concentration cell with cation exchange membrane and AgCl|Ag electrodes.

The cell may be compared to the one described in Section 4.1.1. Electrodes and electrolytes are the same in both cells, but in the present cell the liquid junction is replaced by a cation exchange membrane.

In the liquid junction the concentration gradient of HCl is determined by the concentrations of HCl in the two adjacent electrolytes. In the cation exchange membrane the cation sites, M^-, are filled with H^+ to keep the membrane electrically neutral, and the concentration of HM anywhere in the membrane is determined by the concentration of cation sites. Such a membrane usually contains a large amount of water (sometimes more than half the weight). The concentration of water may vary somewhat through the membrane. If the membrane is not a perfect cation exchanger, it will contain a small amount of the neutral species HCl.

For the cell with a liquid junction the changes in electrolytes I and II with the passage of electric current have been described by eqs. (4.2) and (4.3). Apart from Ag and AgCl at the electrodes, the system consisted of two components, HCl and H_2O, and H_2O was chosen as the frame of reference. The present cells consists of three components, HCl, H_2O and HM (Ag and AgCl at the electrodes). We shall choose HM, the membrane, as the frame of reference for the movements. Since the content of HCl in the membrane is very low (or nil), very little HCl can diffuse through the membrane. In a closed circuit, however, HCl can be transported from one side to the other with the electric current, H^+ via the membrane and Cl^- by the electrode reactions. The net observable result is a transfer of neutral HCl from one side to the other. Water is transported through the membrane by diffusion and together with H^+.

With the membrane as a frame of reference, there are the three fluxes J_{HCl}, J_{H_2O} and j. The three corresponding forces are given the symbols $-\nabla\mu_{HCl}$, $-\nabla\mu_{H_2O}$ and $-\nabla\varphi$. The meaning of these symbols will be discussed below. We have a set of three flux equations (compared to only two for the cell with the liquid junction, see eq. (4.4)):

$$J_1 = J_{HCl} = -L_{11}\nabla\mu_{HCl} - L_{12}\nabla\mu_{H_2O} - L_{13}\nabla\varphi \tag{5.5a}$$

$$J_2 = J_{H_2O} = -L_{21}\nabla\mu_{HCl} - L_{22}\nabla\mu_{H_2O} - L_{23}\nabla\varphi \tag{5.5b}$$

$$j = -L_{31}\nabla\mu_{HCl} - L_{32}\nabla\mu_{H_2O} - L_{33}\nabla\varphi \tag{5.6c}$$

For emf calculations we shall make use of eq. (5.5c) with the condition $j = 0$. We obtain

$$\nabla\varphi = -t_{HCl}\nabla\mu_{HCl} - t_{H_2O}\nabla\mu_{H_2O} \tag{5.6}$$

where $t_{HCl} = L_{31}/L_{33}$, the transference coefficient of HCl, and $t_{H_2O} = L_{32}/L_{33}$, the transference coefficient of H_2O. With gradients in one direction only, we may write,

$$d\varphi = -t_{HCl}d\mu_{HCl} - t_{H_2O}d\mu_{H_2O} \tag{5.7}$$

In order to define the forces $-\nabla\varphi$, $-\nabla\mu_{HCl}$ and $-\nabla\mu_{H_2O}$, or the changes $d\varphi$, $d\mu_{HCl}$ and $d\mu_{H_2O}$ over a distance dx inside the membrane, we may visualize a section of thickness dx cut out of the membrane. The section is placed between two electrolyte solutions, each one of a composition that is in local equilibrium with the membrane section at the phase boundary. Thus the gradients in the section are kept undisturbed. An AgCl|Ag electrode is placed in each solution and the potential measured across this cell when $j \approx 0$ (see Fig. 5.4).

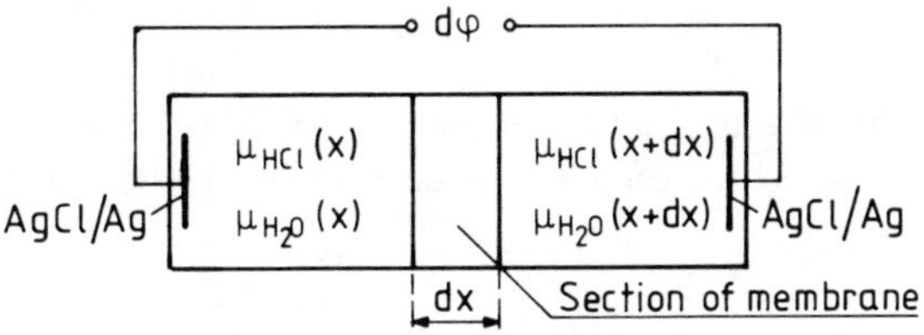

Figure 5.4. Definition of forces in the membrane.

The gradient in electric potential, $d\varphi/dx$, across any section of the membrane is defined as the potential of a cell such as the one pictured in Fig. 5.4. The gradients in chemical potential in the membrane are defined by the differences in chemical potentials, $d\mu_{HCl}/dx = \{\mu_{HCl}(x+dx) - \mu_{HCl}(x)\}/dx$ and $d\mu_{H_2O}/dx = \{\mu_{H_2O}(x+dx) - \mu_{H_2O}(x)\}/dx$. In this way the gradients and forces inside the membrane are defined by external quantities and the membrane is treated as a 'black box'.

A basis for the above definitions is the assumption that the composition anywhere in the membrane has a corresponding equilibrium composition of the HCl–H_2O solution at the external pressure.

The Gibbs–Duhem equation in the solution is used to eliminate $d\mu_{H_2O}$ from eq. (5.7):

$$c_{HCl}d\mu_{HCl} + c_{H_2O}d\mu_{H_2O} = 0 \tag{5.8}$$

and the following expression is obtained:

$$d\varphi = FdE = -\left(t_{HCl} - t_{H_2O}\frac{c_{HCl}}{c_{H_2O}}\right) d\mu_{HCl} \qquad (5.9)$$

The equation may be rearranged and integrated over the membrane from electrolyte I to electrolyte II:

$$\Delta\mu_{HCl} = \mu_{HCl,II} - \mu_{HCl,I} = -F\int_{(I)}^{(II)} dE/\left(t_{HCl} - t_{H_2O}\frac{c_{HCl}}{c_{H_2O}}\right) \qquad (5.10)$$

The value of the integral on the right-hand side of eq. (5.10) can be found by a numerical integration from known values of t_{HCl}, t_{H_2O} and E for different values of c_{HCl}. Thus one can find the chemical potential of HCl in an HCl–H_2O mixture as a function of composition, using the left-hand-side electrolyte as reference state.

Conversely, if $\Delta\mu_{HCl}$, t_{HCl} and t_{H_2O} are known as functions of composition, the emf of the cell can be found by integrating eq. (5.9):

$$FE = -\int_{(I)}^{(II)} \left(t_{HCl} - t_{H_2O}\frac{c_{HCl}}{c_{H_2O}}\right) d\mu_{HCl} \qquad (5.11)$$

The transference coefficient, t_{HCl}, is equal to the transference number, t_{H^+}. The transference coefficient, t_{H_2O}, is the number of moles of water transferred from left to right per faraday transferred. Water molecules are transported to the right with cations and to the left with anions. For the cation exchange membrane t_{H^+} is very close to unity, and we have $t_{H_2O} \approx \alpha_{H^+} t_{H^+}$, where α_{H^+} is the number of water molecules transported with a proton.

The transference number, t_{H^+}, is obtained for any concentration of HCl by a Hittorf experiment using identical electrolytes on both sides of the membrane (cf. Section 4.1.1).

The transference coefficient, t_{H_2O}, can, for example, be obtained from experimental values of emf for a cell with a pressure difference over the membrane. This method will be discussed in Section 5.5.3 (see eq. (5.72)). The contribution to the emf of the present cell, however, from the transport of water is very small, as can be seen from eq. (5.11). For HCl concentrations of the order of magnitude 0.1 kmol m^{-3} and with $t_{H_2O} \approx 1$ (a common value), we have $t_{H_2O}c_{HCl}/c_{H_2O} \approx 1 \times 0.1 \times 1000/(56 \times 10^3) = 0.0018$. When $t_{HCl} = 1$, the contribution from the transport of water is less than 0.2 per cent of the total.

The emf of the cell, E, is measured by means of a potentiometer.

5.3.2 *The HCl–H_2O concentration cells with different types of membranes and different electrodes reversible to one of the ions*

In Section 5.3.1 we studied an HCl–H_2O concentration cell with a cation exchange membrane and AgCl|Ag electrodes. We saw that with a *perfect* cation exchange membrane, $t_{H^+} = 1$, one mole of HCl is transferred per faraday electric charge. The anion-reversible electrodes may be replaced by $H^+|H_2$ electrodes:

$$(\mathrm{Pt})\mathrm{H_2(g,1\ atm)|HCl(aq},c_{\mathrm{I}})|^{\mathrm{C}}|\mathrm{HCl(aq},c_{\mathrm{II}})|\mathrm{H_2(g,1\ atm)(Pt)}$$

In this cell one mole H^+ is produced at the left-hand-side electrode and simultaneously one mole H^+ is consumed at the right-hand-side, per faraday electrons passed from left to right through the outer circuit. With a perfect cation exchange membrane, $t_{H^+} = 1$, the electroneutrality is maintained by the migration of all the produced H^+ from left to right through the membrane. There is no change in the composition of the electrolytes. The flux equations are still eqs. (5.5), but now $J_{HCl} = 0$. The transference coefficient $t = 0$, and eq. (5.7) reduces to

$$d\varphi = -\, t_{H_2O} d\mu_{H_2O} \tag{5.12}$$

The Gibbs–Duhem equation (5.8) is used to replace $d\mu_{H_2O}$ by $d\mu_{HCl}$, and we obtain:

$$d\varphi = F dE = \left(t_{H_2O}\frac{c_{HCl}}{c_{H_2O}}\right) d\mu_{HCl} \tag{5.13}$$

We may introduce $(2RT/c_{HCl})dc_{HCl} = d\mu_{HCl}$ (cf. eq. (4.27a)) and integrate eq. (5.13) over the membrane (neglecting any change in c_{H_2O} from I to II):

$$FE = t_{H_2O} \frac{2RT}{c_{H_2O}} \int_{(I)}^{(II)} dc_{HCl} = 2RT(t_{H_2O}/c_{H_2O})\Delta c_{HCl}$$

When a perfect anion exchange membrane, $t_{Cl^-} = 1$, is used, anion-reversible electrodes will give $J_{HCl} = 0$. The emf of the cell is determined by the transfer of water, as given by eq. (5.12).

When the anion exchange membrane is combined with cation-reversible electrodes eqs. (5.7) and (5.8) give the following:

$$d\varphi = -\left(t_{HCl} - t_{H_2O}\frac{c_{HCl}}{c_{H_2O}}\right) d\mu_{HCl} \tag{5.14}$$

where $t_{HCl} = -t_{Cl^-}$ (cf. eq. (5.9)).

5.3.3 *Influence of pressure gradients*

The simplifying assumption of no pressure differences was made in section 5.3.1. The effect of pressure differences and pressure gradients will now be studied. The same example cell will be used as in Section 5.3.1.

Let us return to Fig. 5.4. The AgCl|Ag electrodes in the electrolytes add the components Ag and AgCl to the system. The transfer of these components is connected to the transfer of charge in a simple way, $J_{Ag} = j$ and $J_{AgCl} = -j$. With a rigid membrane there is a corresponding transfer of volume from left to right equal to $j(V_{Ag} - V_{AgCl}) = j\Delta V_{el}$. Here V_{Ag} and V_{AgCl} are the molar volumes of Ag(s) and AgCl(s) respectively, while ΔV_{el}, the difference between these, is the change in the volume of the electrodes by the passage of one faraday of electrons from left to right through the outer circuit.

The components Ag and AgCl form separate phases, and they are not present in any of the other phases. This means that the gradients in their chemical potentials are equal to zero unless there is a pressure gradient in the system. With a pressure gradient, ∇p, the gradients in chemical potential for the two components are $\nabla\mu_{Ag} = V_{Ag}\nabla p$ and $\nabla\mu_{AgCl} = V_{AgCl}\nabla p$. Since these gradients are forces related to the flux j, the contribution to the gradient in the *observed electric potential*, $\nabla\varphi_{obs}$, from these components is equal to $-\Delta V_{el}\nabla p$. Hence the gradient in electric potential occurring in eqs. (5.5), $\nabla\varphi$, is the sum of $\nabla\varphi_{obs}$ and the term $\Delta V_{el}\nabla p$:

$$\nabla\varphi = \nabla\varphi_{obs} + \Delta V_{el}\nabla p \tag{5.15}$$

The gradients in chemical potential anywhere inside the membrane are defined by gradients in the solutions in local equilibrium with the membrane. For an isothermal system where $\nabla\mu_i$ includes concentration and pressure gradients, we have

$$\nabla\mu_i = \nabla\mu_i(c) + V_i\nabla p \tag{5.16}$$

where $\nabla\mu_i(c)$ is a function of concentration gradients only, V_i is the partial molar volume of component i and ∇p is the gradient in pressure. Since $\Sigma_i n_i V_i = V$, we have the Gibbs–Duhem equation at constant temperature:

$$\Sigma_i n_i \nabla\mu_i - V\nabla p = \Sigma_i n_i \nabla\mu_i(c) = 0 \tag{5.17}$$

When energy contributions due to change in dipole orientation are negligible, eqs. (5.16) and (5.17) do not contain gradients in electric potential. de Groot and Mazur[3] discuss these energy contributions.

According to the Gibbs–Duhem equation all except one of the gradients can be varied independently. If the pressure gradient is allowed to adjust itself, all the gradients in chemical potential can be varied independently. For the two-component system of Section 5.3.1, both $\nabla\mu_{HCl}$ and $\nabla\mu_{H_2O}$ can be varied independently. The gradients in $\nabla\mu_i(c)$, however, are interdependent; if one is changed, the other one changes also.

5.4 *Electrokinetic Phenomena and Osmosis*

Several electrokinetic phenomena may occur in a cell where two electrolytes are separated by a membrane. The more important ones are the *streaming potential*, the electric potential created by a pressure difference, *electro-osmotic flux*, the volume flux connected to an electric current, *electro-osmotic pressure*, the pressure difference created by an electric current when volume flux is prevented, and *streaming current,* the electric current created by the volume flux.

The phenomenon of osmosis may occur when two solutions of different composition are separated by a semipermeable membrane. The *osmotic pressure* is the pressure difference between the two electrolytes needed to arrest the diffusion of solvent through the membrane. The *Donnan equilibrium* and the *reflection coefficient* are important concepts related to osmotic phenomena.

5.4.1 *Electrokinetic phenomena*

In order to study electrokinetic phenomena and their interrelations, we shall use an HCl–H_2O cell with a cation exchange membrane and AgCl|Ag electrodes. Both electrolytes have the same composition, but there is a pressure difference between the two electrolytes (see Fig. 5.5).

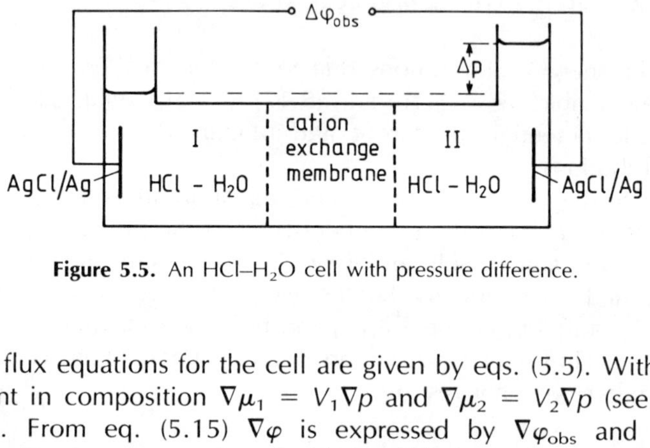

Figure 5.5. An HCl–H_2O cell with pressure difference.

The flux equations for the cell are given by eqs. (5.5). With no gradient in composition $\nabla\mu_1 = V_1\nabla p$ and $\nabla\mu_2 = V_2\nabla p$ (see eq. (5.16)). From eq. (5.15) $\nabla\varphi$ is expressed by $\nabla\varphi_{\text{obs}}$ and ∇p. Introducing these expressions for $\nabla\mu_i$ and $\nabla\varphi$ in the flux equations, we obtain

$$J_1 = -(L_{11}V_1 + L_{12}V_2 + L_{13}\Delta V_{\text{el}})\nabla p - L_{13}\nabla\varphi_{\text{obs}} \tag{5.18a}$$

$$J_2 = -(L_{21}V_1 + L_{22}V_2 + L_{23}\Delta V_{\text{el}})\nabla p - L_{23}\nabla\varphi_{\text{obs}} \tag{5.18b}$$

$$j = -(L_{31}V_1 + L_{32}V_2 + L_{33}\Delta V_{\text{el}})\nabla p - L_{33}\nabla\varphi_{\text{obs}} \tag{5.18c}$$

The *volume flux* involved in the electrokinetic phenomena is defined as

$$J_V = J_1 V_1 + J_2 V_2 + j\Delta V_{el} \tag{5.19}$$

At constant temperature the coefficients L_{ij} are functions of composition and pressure. With electrolytes of the same composition we assume that there is no gradient in composition in the membrane. For moderate pressure differences (order of magnitude 1 atm), the pressure effect on the equilibrium between membrane and solution will not cause any significant gradient in composition. Furthermore, the changes in L_{ij} with pressure are negligible for moderate pressure differences. Hence all L_{ij} may be considered to be constant throughout the membrane. When the experiments are carried out in the stationary state, the fluxes are constant. With constant coefficients and constant fluxes, the forces are also constant through the membrane, $\nabla p = \Delta p/\Delta x$ and $\nabla \varphi_{obs} = \Delta\varphi_{obs}/\Delta x$. Here Δx is the thickness of the membrane, Δp is the pressure difference between electrolytes II and I, and $\Delta\varphi_{obs}$ is the electric potential of the cell. Experimental values can be obtained for all of them.

We may introduce Δp and $\Delta\varphi_{obs}$ in eqs. (5.18), normalize Δx to unity and consider the fluxes over the whole cross-section. At the same time the equations may be simplified by using the following notation:

$$\mathcal{L}_i = (L_{i1}V_1 + L_{i2}V_2 + L_{i3}\Delta V_{el}); \quad (i = 1,2,3) \tag{5.20}$$

Thus we obtain

$$J_1 = -\mathcal{L}_1 \Delta p - L_{13}\Delta\varphi_{obs} \tag{5.21a}$$

$$J_2 = -\mathcal{L}_2 \Delta p - L_{23}\Delta\varphi_{obs} \tag{5.21b}$$

$$I = -\mathcal{L}_3 \Delta p - L_{33}\Delta\varphi_{obs} \tag{5.21c}$$

The coefficients in these equations should be multiplied by the thickness of the membrane to compare them with the coefficients of eqs. (5.18). Corresponding to eqs. (5.21) the expression for the volume flux is

$$J_V = -(\mathcal{L}_1 V_1 + \mathcal{L}_2 V_2 + \mathcal{L}_3 \Delta V_{el})\Delta p - \mathcal{L}_3 \Delta\varphi_{obs} \tag{5.22}$$

The *streaming potential* of a cell is defined as the observed electric potential across the cell divided by the pressure difference across the cell when the current is equal to zero. Applying eq. (5.21c) for $I = 0$, the following expression is obtained for the streaming potential:

$$\left[\frac{\Delta\varphi_{obs}}{\Delta p}\right]_{I=0,\Delta\mu_i(c)=0} = -\frac{\mathcal{L}_3}{L_{33}} = -(t_1V_1 + t_2V_2 + \Delta V_{el}) \quad (5.23)$$

The *electro-osmotic flux* across a cell is defined as the volume flux across the cell divided by the electric current, when there is no pressure difference. Dividing eq. (5.22) by eq. (5.21c) for $\Delta p = 0$, the following expression is obtained for the electro-osmotic flux:

$$\left[\frac{J_V}{I}\right]_{\Delta p=0,\Delta\mu_i(c)=0} = +\frac{\mathcal{L}_3}{L_{33}} = t_1V_1 + t_2V_2 + \Delta V_{el} \quad (5.24)$$

Note that

$$\left[\frac{\Delta\varphi_{obs}}{\Delta p}\right]_{I=0,\Delta\mu_i(c)=0} = -\left[\frac{J_V}{I}\right]_{\Delta p=0,\Delta\mu_i(c)=0}$$

The *electro-osmotic pressure* of a cell is defined as the pressure difference across the cell divided by the electric potential of the cell, when the volume flux is equal to zero. Applying eq. (5.22) for $J_V = 0$, we obtain

$$\left[\frac{\Delta p}{\Delta\varphi_{obs}}\right]_{J_V=0,\Delta\mu_i(c)=0} = -\frac{\mathcal{L}_3}{\mathcal{L}_1V_1 + \mathcal{L}_2V_2 + \mathcal{L}_3\Delta V_{el}} \quad (5.25)$$

The *streaming current* in a cell is defined as the electric current in the cell divided by the volume flux when the electric potential across the cell is equal to zero (zero electric potential is obtained by short-circuiting the electrodes). Dividing eq. (5.21c) by eq. (5.22) for $\Delta\varphi_{obs} = 0$, we obtain

$$\left[\frac{I}{J_V}\right]_{\Delta\varphi_{obs}=0,\Delta\mu_i(c)=0} = \frac{\mathcal{L}_3}{\mathcal{L}_1V_1 + \mathcal{L}_2V_2 + \mathcal{L}_3\Delta V_{el}} \quad (5.26)$$

Note that

$$\left[\frac{\Delta p}{\Delta \varphi_{obs}}\right]_{J_V=0,\Delta\mu_i(c)=0} = -\left[\frac{I}{J_V}\right]_{\Delta\varphi_{obs}=0,\Delta\mu_i(c)=0}$$

5.4.2 *Osmosis*

Next we shall consider *osmosis* in a system of two solutions separated by a membrane. The system does not contain electrodes, no electric energy is supplied to the system and no net charge is transported through the system. The forces of the system are gradients in chemical potential including concentration gradients and pressure gradients. Systems of this kind are very common in biology. The membrane may be permeable to the solvent (water) only, or it may have, in addition, a selective permeability to different solutes. Also, transports across the membrane may be coupled. Thermodynamic relations and transport equations will be derived for some simple systems of this kind.

For a system of two components in solutions separated by a membrane, the flux equations may be written with diffusion coefficients:

$$J_1 = -l_{11}\nabla\mu_1 - l_{12}\nabla\mu_2 \tag{5.27a}$$

$$J_2 = -l_{21}\nabla\mu_1 - l_{22}\nabla\mu_2 \tag{5.27b}$$

where components 1 and 2 are solvent and solute respectively.

If the membrane is permeable to component 1 only, and completely impermeable to component 2, then $l_{22} = l_{12} = l_{21} = 0$. This reduces eqs. (5.27) and with eq. (5.16) we obtain

$$J_1 = -l_{11}\nabla\mu_1 = -l_{11}(\nabla\mu_1(c) + V_1\nabla p) \tag{5.28}$$

The *osmotic pressure*, Π, is commonly defined as the pressure difference over such a semipermeable membrane that will make J_1 equal to zero. When $J_1 = 0$, we obtain from eq. (5.28)

$$\nabla p = -\frac{\nabla\mu_1(c)}{V_1} \tag{5.29}$$

We can multiply and divide by c_1 on the righthand side of eq. (5.29). For a sufficiently dilute solution we may consider $c_1V_1 \approx 1$, which gives,

$$\nabla p = -c_1\nabla\mu_1(c) \tag{5.30}$$

From the Gibbs–Duhem equation applied to solutions in equilibrium with the membrane (cf. Fig. 5.4), we have $c_2\nabla\mu_2(c) = -c_1\nabla\mu_1(c)$, and the gradient in pressure can be expressed by the concentration and the gradient in chemical potential of the solute

$$\nabla p = c_2\nabla\mu_2(c) \tag{5.31}$$

For an ideal solution of a non-electrolyte $\nabla\mu_2(c)$ may be replaced by $RT\nabla c_2/c_2$ (see eq. (4.27b)). Integrating over the membrane, the osmotic pressure is obtained:

$$\Delta p = \Pi = RT\Delta c_2 \tag{5.32}$$

This is the *van't Hoff equation*[4] of osmotic pressure for a solution of a non-electrolyte.

The solute, component 2, may be a strong electrolyte, NaR, where the large anion R^- cannot permeate the membrane because of size. Since electroneutrality is preserved throughout the system, the smaller cation, Na^+, is also prevented from diffusion through the membrane. When the solution is ideal, $\nabla\mu_2(c) = RT\nabla\ln(c_{Na^+}c_{R^-})$ and, since $c_{Na^+} = c_{R^-} = c_2$, we may replace $\nabla\mu_2(c)$ by $2RT\nabla c_2/c_2$ (see eq. (4.27a)). Integrating over the membrane we obtain

$$\Delta p = \Pi = 2RT\Delta c_2 \tag{5.33}$$

which is the equation of osmotic pressure for a solution of a strong electrolyte.

The Donnan equilibrium and osmotic pressure. We shall consider a system of two strong electrolytes, NaCl and NaR, in aqueous solution. Water and the smaller ions Na^+ and Cl^- can diffuse through the pores of the membrane, while the larger anion R^- cannot permeate the membrane. The R^- is on one side of the membrane only, in electrolyte solution II (see Fig. 5.6).

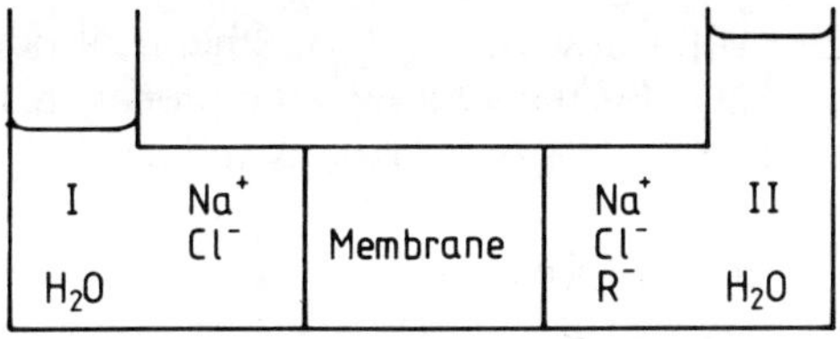

Figure 5.6. The Donnan equilibrium.

If Na^+ and Cl^- diffuse through the membrane easily, an equilibrium for NaCl will quickly be established across the membrane, $\mu_{NaCl,I} = \mu_{NaCl,II}$. When there is no difference in pressure between I and II, however, there can be no equilibrium for water, $\mu_{H_2O,I} > \mu_{H_2O,II}$. Thus water will continue to diffuse from compartment I to compartment II. The equilibrium for NaCl across the membrane is upheld by a continuous diffusion of NaCl with the water in the same direction. Thus, sooner or later, compartment I will be emptied. In order to establish a true equilibrium across the membrane, we must apply a pressure to compartment II, and this pressure is defined as equal to the osmotic pressure of the solution in compartment II.

The equilibrium for NaCl across the membrane, $\mu_{NaCl,I} = \mu_{NaCl,II}$, is known as the *Donnan equilibrium*.[5] The concentrations in compartment I are $c_{Na^+,I} = c_{Cl^-,I} = c$, while in compartment II they are $c_{Na^+} = c_{R^-} + c_{Cl^-}$. For ideal solutions at equal pressures $\mu_{NaCl,I} = RT\ln c^2$, while $\mu_{NaCl,II} = RT\ln (c_{Na^+}c_{Cl^-})$. Thus the Donnan equilibrium can be written as follows:

$$c^2 = (c_{R^-} + c_{Cl^-})c_{Cl^-} \tag{5.34}$$

The flux equations for the system NaCl, NaR, H_2O and membrane are

$$J_{H_2O} = J_1 = -l_{11}\nabla\mu_1 - l_{12}\nabla\mu_2 - l_{13}\nabla\mu_3$$
$$J_{NaR} = J_2 = -l_{21}\nabla\mu_1 - l_{22}\nabla\mu_2 - l_{23}\nabla\mu_3$$
$$J_{NaCl} = J_3 = -l_{31}\nabla\mu_1 - l_{32}\nabla\mu_2 - l_{33}\nabla\mu_3$$

where $\nabla\mu_1 = \nabla\mu_{H_2O}$, $\nabla\mu_2 = \nabla\mu_{NaR}$ and $\nabla\mu_3 = \nabla\mu_{NaCl}$.

Since $J_{NaR} = 0$ for any value of the chemical potentials, $l_{21} = l_{22} = l_{23} = 0$, and by the Onsager reciprocal relations $l_{12} = 0$ and $l_{32} = 0$. Thus the set of equations reduces to

$$J_1 = -l_{11}\nabla\mu_1 - l_{13}\nabla\mu_3 \tag{5.35a}$$

$$J_3 = -l_{31}\nabla\mu_1 - l_{33}\nabla\mu_3 \tag{5.35b}$$

A pressure can be applied to compartment II to stop the fluxes. The Donnan equilibrium, $\Delta\mu_{NaCl} = \Delta\mu_3 = 0$, will be re-established, and then $\nabla\mu_3 = 0$ over the membrane. When the fluxes have been stopped, $\nabla\mu_1 = \nabla\mu_1(c) + V_1\nabla p = 0$, which gives $\nabla p = -\nabla\mu_1(c)/V_1$ (cf. eq. (5.29)). By the Gibbs–Duhem equation $-\nabla\mu_1(c)$ can be replaced by $(c_2/c_1)\nabla\mu_2(c) + (c_3/c_1)\nabla\mu_3(c)$, which gives

$$c_1V_1\nabla p = c_2\nabla\mu_2(c) + c_3\nabla\mu_3(c) \tag{5.36}$$

When integrating eq. (5.36), we shall arrange the experiment such that the gradients are in one direction only, $\nabla p = dp/dx$ and $\nabla\mu = d\mu/dx$. We shall also assume that at any distance inside the membrane, we can find solutions of NaR and NaCl that are in equilibrium over a narrow section of the membrane of thickness dx. We may develop each term on the right-hand side of eq. (5.36) separately, assuming an ideal solution.

$$c_2d\mu_2(c) = RTc_{R^-}d\ln(c_{Na^+}\cdot c_{R^-}) = \frac{RT}{c_{Na^+}}(c_{Na^+}dc_{R^-} + c_{R^-}dc_{Na^+})$$

$$c_3d\mu_3(c) = RTc_{Cl^-}d\ln(c_{Na^+}\cdot c_{Cl^-}) = \frac{RT}{c_{Na^+}}(c_{Na^+}dc_{Cl^-} + c_{Cl^-}dc_{Na^+})$$

The terms may be added up, remembering that electroneutrality requires that $c_{Na^+} = c_{R^-} + c_{Cl^-}$ and that $dc_{Na^+} = dc_{R^-} + dc_{Cl^-}$:

$$c_2d\mu_2(c) + c_3d\mu_3(c) = 2RTdc_{Na^+}$$

In a dilute solution we have $c_1V_1 = c_{H_2O}V_{H_2O} \approx 1$ and thus, by applying eq. (5.36):

$$dp = 2RTdc_{Na^+} \tag{5.37}$$

The osmotic pressure is

$$\Delta p = \Pi = 2RT \int_{c_{Na^+}=c}^{c_{Na^+}=c_{R^-}+c_{Cl^-}} dc_{Na^+} = 2RT(c_{R^-} + c_{Cl^-} - c) \tag{5.38}$$

It can be shown that the shift in the Donnan equilibrium is negligible for reasonable osmotic pressures. To a good approximation eq. (5.34) can be used to obtain an expression for c_{Cl^-} by means of c_{R^-} and c and the osmotic pressure can be expressed by c_{R^-} and c:

$$\Pi = RT\{\sqrt{4c^2 - c_{R^-}^2} + c_{R^-} - 2c\}$$

Two extreme cases may be considered: (i) when $c = 0$, then $\Pi = 2RTc_{R^-}$ (cf. eq. (5.33)) and (ii) when $c \gg c_{R^-}$, the approximation $\sqrt{1 + x} \approx 1 + \frac{1}{2}x$ may be used and $\Pi = RT(1 + c_{R^-}/4c)\, c_{R^-} \approx RTc_{R^-}$ is obtained (cf. eq. (5.32)).

For an intermediate case where c and c_{R^-} are of the same order of magnitude, it may be convenient to arrange the experiment with a large reservoir of solution in compartment I, so that the change in c by the diffusion of NaCl and water can be neglected.

The reflection coefficient. So far we have only considered the osmotic pressure when the membrane is completely impermeable to a solute. The osmotic pressure is then equal to the pressure that must be applied to stop the flux of the solvent. When the membrane is not completely impermeable to a solute, the flux of this component must also be considered. We shall assume no passage of bulk solution through the membrane.

In an osmotic experiment the hydrostatic pressure is measured when there is no volume flux, $J_V = 0$. When there are two components that can both diffuse through the membrane

$$J_V = J_1V_1 + J_2V_2 \tag{5.39}$$

and when $J_V = 0$, there may still be fluxes through the membrane, $J_1V_1 = -J_2V_2$.

The two components have different velocities, $J_1 = c_1v_1$ and $J_2 = c_2v_2$, where v_1 and v_2 are the respective velocities of components 1 and 2. The difference in velocity, $v_2 - v_1$, is called J_D, the *diffusion flux*:

$$J_D = J_2/c_2 - J_1/c_1 \tag{5.40}$$

(The dimension for J_1 and J_2 is mol $m^{-2}s^{-1}$. The dimension for partial molar volume is m^3 mol^{-1} and for concentration mol m^{-3}. Thus J_V and J_D both have the dimension m s^{-1}.)

It is convenient to use the fluxes J_V and J_D in the description of osmotic phenomena in membranes that are permeable to both components. This is commonly done in biophysics, and the selectivity of a membrane is characterized by the *reflection coefficient*, σ. The reflection coefficient was introduced by Staverman.[6, 7] The present treatment is mainly based on the work of Katchalsky and coworkers.[8, 9]

The dissipation function was defined by eq. (2.26) as a sum of forces – flux products. Forces and conjugate fluxes may be defined in different ways:

$$T\Theta = \Sigma_j X_j J_j = \Sigma_j X_j' J_j' \tag{5.41}$$

We have seen that when we have both a pressure gradient and a composition gradient in a system, the chemical potential contains a concentration term and a pressure term (see eq. (5.16)). The gradients in the chemical potentials are related to one another by the Gibbs–Duhem equation (see eq. (5.17)). The two forces, $-\nabla p$ and $-c_2\nabla\mu_2(c)$ will be introduced into the flux equations instead of the forces $-\nabla\mu_1$ and $-\nabla\mu_2$. (The dimension for ∇p and for $c_2\nabla\mu_2(c)$ is J m^{-4}). The transformation from the old fluxes and forces to the new ones is

$$\begin{aligned} T\Theta &= -J_1\nabla\mu_1 - J_2\nabla\mu_2 = -J_1(\nabla\mu_1(c) + V_1\nabla p) - J_2(\nabla\mu_2(c) + V_2\nabla p) \\ &= -(J_1V_1 + J_2V_2)\nabla p - (J_2/c_2 - J_1/c_1)c_2\nabla\mu_2(c) \end{aligned}$$

and by eqs. (5.39) and (5.40)

$$T\Theta = -J_V\nabla p - J_D c_2\nabla\mu_2(c) \tag{5.42}$$

The corresponding flux equations are

$$J_V = -L_p \nabla p - L_{PD} c_2 \nabla \mu_2(c) \tag{5.43a}$$

$$J_D = -L_{DP} \nabla p - L_D c_2 \nabla \mu_2(c) \tag{5.43b}$$

where $L_{PD} = L_{DP}$ according to the Onsager reciprocal relations.

Again we shall arrange the experiment with gradients in the x-direction only and measure differences across the membrane. We shall measure the hydrostatic pressure difference across the membrane that makes J_v equal to zero. In an ideal solution of a non-electrolyte $c_2 \nabla \mu_2(c) = RT \nabla c_2$, and from eq. (5.43a) we obtain

$$\Delta p = -\frac{L_{PD}}{L_P} RT \Delta c_2; \quad (J_V = 0) \tag{5.44}$$

The ratio $-L_{PD}/L_P$ is defined as the *reflection coefficient*:

$$\sigma = -\frac{L_{PD}}{L_P} = \left[\frac{\Delta p}{RT \Delta c_2}\right]_{J_V=0} \tag{5.45}$$

Both Δp and Δc_2 will decrease over time, since the continuous fluxes, J_1 and J_2, will level out differences across the membrane.

The reflection coefficient may be interpreted as the ratio between the experimental pressure difference and the osmotic pressure that would have been obtained if the membrane were completely impermeable to component 2. Then, according to eq. (5.32) $\Pi = RT \Delta c_2$ and thus $\sigma = \Delta p/\Pi$. To study the interpretation of σ more closely, we may look at another way of obtaining it. When both J_V and J_D are measured as functions of Δp for $\Delta \mu_2(c) = 0$, we have

$$\left[\frac{J_D}{J_V}\right]_{\Delta \mu_2(c)=0} = \frac{L_{DP}}{L_P} = \frac{L_{PD}}{L_P} = -\sigma \tag{5.46}$$

Using eqs. (5.39) and (5.40) we obtain

$$\left[\frac{J_2/c_2 - J_1/c_1}{J_1 V_1 + J_2 V_2}\right]_{\Delta \mu_2(c)=0} = -\sigma \tag{5.47}$$

If the membrane is permeable to component 1 only and completely impermeable to component 2, then $J_2 = 0$, which gives

$$\sigma = \frac{1}{c_1 V_1}; \qquad (J_2 = 0) \tag{5.48}$$

When component 1 is the solvent and the solution is very dilute, $c_1 V_1 \approx 1$ and the reflection coefficient is approximately equal to unity ($\sigma \approx 1$).

When the velocities of components 1 and 2 are equal ($J_2/c_2 = J_1/c_1$), the value of σ is equal to zero ($\sigma = 0$). The membrane is equally permeable to both components.

A value of σ between one and zero means that the membrane is 'leaking'. The solute, component 2, migrates through the membrane, but at a slower rate than the solvent, component 1.

Negative anomalous osmosis may be found in some systems. In such systems the pressure must be applied in the opposite direction, i.e. the more dilute solution must be exposed to the higher pressure, in order to keep J_V equal to zero. This means that $J_2/c_2 > J_1/c_1$, when there is no concentration gradient, and $\sigma < 0$ (see eq. (5.47)). In the extreme case when the passage of the solvent through the membrane is completely blocked, $J_1 = 0$.

Of particular interest are systems where aqueous solutions of electrolytes are separated by a membrane permeable to ions (positive and negative) while the permeability to neutral species, such as water, is very low. Water may be transported through the membrane with the ions, i.e. there is a *coupling* between the transports of the two components, electrolyte and water. Since usually $c_2 \ll c_1$, we may have $J_2/c_2 > J_1/c_1$ and $\sigma < 0$, even when several water molecules are transported with each ion.

Systems of several components. The solutions may contain more than one solute. The composition of the system must be described by the minimum number of components in the thermodynamic sense, to ensure that the changes in chemical potentials lead to independent forces (see Section 3.3). The electroneutrality of both solutions leads to a complete coupling between the transports of positive and negative ions.

When there are n components in the solution, the dissipation function may be written as

$$T\Theta = -\Sigma_i J_i \nabla\mu_i \tag{5.49}$$

Component 1 may be considered as the solvent, and by a treatment similar to that for two components, an expression can be obtained for $T\Theta$ with $-\nabla p$ and $-c_i\nabla\mu_i(c)$ as forces (cf. eq. (5.42)):

$$T\Theta = -J_V\nabla p - \sum_{i=2}^{n} (J_i/c_i - J_1/c_1)c_i\nabla\mu_i(c) \tag{5.50}$$

The corresponding flux equations for the volume flux, J_V, can be written as

$$J_V = -L_p\{\nabla p - \sum_{i=2}^{n} \sigma_i c_i \nabla\mu_i(c)\} \tag{5.51}$$

where σ_i is the reflection coefficient for component i.

5.5 *A Concentration Cell with Electrolytes Containing Two Salts of the Same Anion, Separated by a Cation Exchange Membrane*

In this section we shall study the coefficients of the flux equations, how they can be interpreted, and what types of experiments can give the values of the coefficients. We shall not consider any temperature gradient, as the system is *isothermal*. We shall consider *gradients in chemical and electric potentials*, and *pressure gradients*.

As an example a system will be chosen where both electrolytes are aqueous solutions of HCl and NaCl, i.e. there are two different cations with a common anion. The electrolytes are separated by a rigid cation exchange membrane, and the AgCl|Ag electrodes are reversible to the common anion:

$$\mathrm{Ag(s)|AgCl(s)|HCl(aq,}c_{1,\mathrm{I}}\mathrm{),\ NaCl(aq,}c_{2,\mathrm{I}}\mathrm{)|^{C}|}$$
$$\mathrm{HCl(aq,}c_{1,\mathrm{II}}\mathrm{),\ NaCl(aq,}c_{2,\mathrm{II}}\mathrm{)|AgCl(s)|Ag(s)}$$

The cell is illustrated schematically in Fig. 5.7.

As in earlier examples, each electrolyte solution will be assumed to be of constant properties. Gradients occur only in the membrane.

In the thermodynamic sense the system contains the four components, HCl, NaCl, H_2O and HM, where HM denotes the membrane in the hydrogen form. The membrane in the sodium form, NaM, is not a component in the thermodynamic sense.

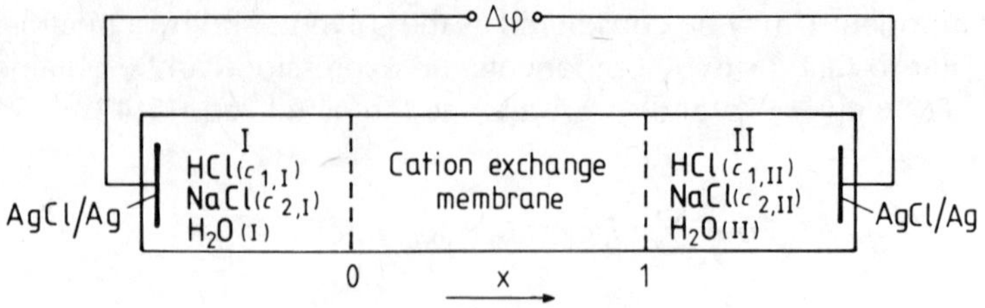

Figure 5.7. A cation exchange membrane separating two solutions of HCl and NaCl. The electrodes are reversible to the Cl^- ion. Reproduced by permission of *Acta Chemica Scandinavica* from Ref. 10.

Assuming local equilibrium inside the membrane, and that the composition anywhere in the membrane corresponds to an equilibrium composition of HCl and NaCl in aqueous solution, NaM can be expressed by the other components in the equilibrium equation:

$$\mathrm{NaM} + \mathrm{HCl} \rightleftarrows \mathrm{NaCl} + \mathrm{HM} \tag{5.52}$$

Thus it is always possible to avoid the component NaM when describing a membrane consisting of x mol HM and $(1 - x)$ mol NaM. The membrane may be considered to contain

$$\begin{aligned} & x \text{ mol HM} + (1 - x) \text{ mol (HM + NaCl} - \text{HCl)} \\ & = 1 \text{ mol HM} + (1 - x) \text{ mol NaCl} - (1 - x) \text{ mol HCl} \end{aligned} \tag{5.53}$$

Notice that the amount of HM is always 1 mol (1 equivalent).

5.5.1 *The set of flux equations*

When diffusion takes place across a membrane while a current passes through the system, the changes taking place are described by fluxes of the neutral components, J_{HCl}, J_{NaCl} and J_{H_2O} and the electric current density, j. The membrane in the hydrogen form is chosen as the frame of reference. This means that the flux of HM, J_{HM}, is chosen as equal to zero, and the other fluxes are given relative to J_{HM}. The reference, HM, is constant for all membrane compositions (see eq. (5.53)).

The forces, $-\nabla\mu_1$, $-\nabla\mu_2$ and $-\nabla\mu_3$, are given by the gradients in chemical potentials, and $-\nabla\varphi$ by the gradient in electric potential. The fluxes are related to the forces by the phenomenological coefficients, L_{ij}, and we have the following flux equations for the system,

$$J_{HCL} = J_1 = -L_{11}\nabla\mu_1 - L_{12}\nabla\mu_2 - L_{13}\nabla\mu_3 - L_{14}\nabla\varphi \quad (5.54a)$$

$$J_{NaCl} = J_2 = -L_{21}\nabla\mu_1 - L_{22}\nabla\mu_2 - L_{23}\nabla\mu_3 - L_{24}\nabla\varphi \quad (5.54b)$$

$$J_{H_2O} = J_3 = -L_{31}\nabla\mu_1 - L_{32}\nabla\mu_2 - L_{33}\nabla\mu_3 - L_{34}\nabla\varphi \quad (5.54c)$$

$$j = J_4 = -L_{41}\nabla\mu_1 - L_{42}\nabla\mu_2 - L_{43}\nabla\mu_3 - L_{44}\nabla\varphi \quad (5.54d)$$

In order to define the potential gradients in eqs. (5.54), we may visualize a section of thickness dx cut out of the membrane, as described in Section 5.3. The section is placed between two electrolyte solutions, each one of a composition that is in local equilibrium with the membrane section at the phase boundary; the gradients in the section are thus kept undisturbed. An AgCl|Ag electrode is placed in each electrolyte and the electrodes are connected by the outer circuit. This subcell is illustrated in Fig. 5.8.

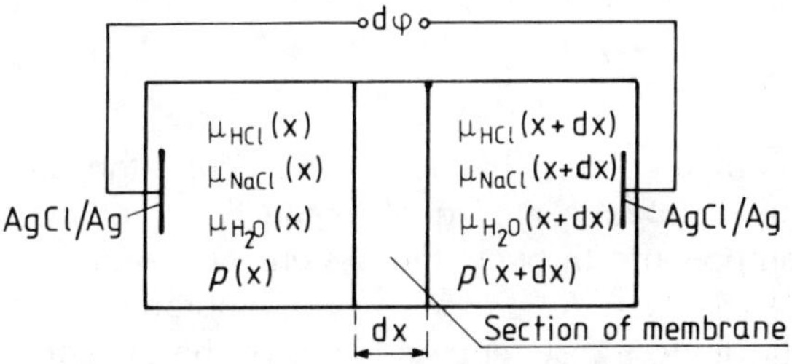

Figure 5.8. A section of the membrane in equilibrium with electrolyte solutions.

As in Section 5.3, the potential gradients inside the membrane are defined by the differences in the potentials between the solutions on the two sides of the membrane. There are no simplifying assumptions about constant pressure, and the forces are given by eqs. (5.15) and (5.16), or with gradients in the x-direction only, $d\varphi = d\varphi_{obs} + \Delta V_{el}dp$ and $d\mu_i = d\mu_i(c) + V_i dp$.

The chemical potentials of HM and NaM are expressed by the chemical potentials of HCl, NaCl and H_2O, and the pressure. The chemical potential of NaM is eliminated by means of the membrane exchange equilibrium restriction (eqs. (5.52) and (5.53)). The chemical potential of HM, the reference, is related to the chemical potentials of HCl, NaCl and H_2O by the Gibbs–Duhem equation (see eq. (5.17)). This means that $\nabla\mu_1$, $\nabla\mu_2$ and $\nabla\mu_3$ of eqs. (5.54) are well-defined functions anywhere in the membrane (even when the content of Cl^- in the membrane is negligible).[10]

5.5.2 *Restrictions on the phenomenological coefficients*

According to the Onsager reciprocal relations, $L_{ij} = L_{ji}$. Additional relations between the phenomenological coefficients will be derived below.

The coefficient L_{44} is related to the *electric conductivity* of the membrane, $L_{44} = \kappa/F^2$ (cf. eq. (4.6)). The coefficients L_{14}, L_{24} and L_{34} may be identified by means of a Hittorf experiment (see Section 4.1.1). In the Hittorf experiment all gradients in chemical potential are equal to zero, and the following information is obtained:

$$(J_1/j)_{\nabla\mu_i=0} = L_{14}/L_{44} = t_1 \tag{5.55a}$$

$$(J_2/j)_{\nabla\mu_i=0} = L_{24}/L_{44} = t_2 \tag{5.55b}$$

$$(J_3/j)_{\nabla\mu_i=0} = L_{34}/L_{44} = t_3 \tag{5.55c}$$

With electrodes reversible to the Cl^- ion, the cation fluxes through the membrane are equal to the fluxes of the respective neutral components through the system, $J_1 = J_{HCl} = J_{H^+}$ (in membrane) and $J_2 = J_{NaCl} = J_{Na^+}$ (in membrane). Therefore t_1 is equal to the ionic transference number of the H^+ ion, $t_1 = t_{H^+}$, and similarly $t_2 = t_{Na^+}$. The t_3 is the transference coefficient of neutral water; it does not represent any transfer of charge. For a cation exchange membrane $t_1 + t_2 = 1$, or stated in an equivalent way:

$$L_{14} + L_{24} = L_{44} \tag{5.56}$$

The current density, j, can be expressed by the fluxes J_1 and J_2:

$$j = J_1 + J_2 \tag{5.57}$$

The gradient in electric potential, $\nabla\varphi$, can be eliminated in the mass flux equations (eqs. (5.54a–c)). Equation (5.54d) is solved with respect to $\nabla\varphi$ and the ratios of phenomenological coefficients are replaced by t_1, t_2 and t_3 according to eqs. (5.55) and the Onsager reciprocal relations,

$$\nabla\varphi = -t_1\nabla\mu_1 - t_2\nabla\mu_2 - t_3\nabla\mu_3 - j/L_{44} \tag{5.58}$$

Substituting this expression in eqs. (5.54a–c) gives,

$$J_1 = -(L_{11}-t_1L_{14})\nabla\mu_1 - (L_{12}-t_2L_{14})\nabla\mu_2 - (L_{13}-t_3L_{14})\nabla\mu_3 + t_1 j$$
$$J_2 = -(L_{21}-t_1L_{24})\nabla\mu_1 - (L_{22}-t_2L_{24})\nabla\mu_2 - (L_{23}-t_3L_{24})\nabla\mu_3 + t_2 j$$
$$J_3 = -(L_{31}-t_1L_{34})\nabla\mu_1 - (L_{32}-t_2L_{34})\nabla\mu_2 - (L_{33}-t_3L_{34})\nabla\mu_3 + t_3 j$$

For convenience the equations may be written in the following abbreviated form:

$$J_1 = -l_{11}\nabla\mu_1 - l_{12}\nabla\mu_2 - l_{13}\nabla\mu_3 + t_1 j \tag{5.59a}$$

$$J_2 = -l_{21}\nabla\mu_1 - l_{22}\nabla\mu_2 - l_{23}\nabla\mu_3 + t_2 j \tag{5.59b}$$

$$J_3 = -l_{31}\nabla\mu_1 - l_{32}\nabla\mu_2 - l_{33}\nabla\mu_3 + t_3 j \tag{5.59c}$$

The equations may be compared to eq. (4.9). When the current density equals zero, eqs. (5.59) represent a homogeneous set of equations with a symmetric matrix. Compared to eqs. (5.54) the number of equations is reduced by one, and the number of coefficients is reduced from 16 to 9. The *diffusion coefficients*, l_{ij} are independent of the kind of electrodes used for measuring $\nabla\varphi$.

Equation (5.60) gives the relation between diffusion coefficients, obtained from pure diffusion experiments, and the phenomenological coefficients, obtained from experiments involving mixed diffusion and electric transport across a cation exchange membrane between anion-reversible electrodes (cf. eq. (4.10)):

$$l_{ij} = L_{ij} - \frac{L_{4j}L_{i4}}{L_{44}}; \quad (i,j = 1,2,3) \tag{5.60}$$

When the Onsager reciprocal relations $L_{ij} = L_{ji}$ are valid, the relations $l_{ij} = l_{ji}$ are also valid. This can be seen from eq. (5.60). We shall apply Gibbs' phase rule to the membrane – electrolyte system in order to show that the forces $-\nabla\mu_1$, $-\nabla\mu_2$ and $-\nabla\mu_3$ of eqs. (5.59) are independent of one another.[10] The fluxes, however, may be interdependent, and in our case we have $J_1 + J_2 = j$.

Local equilibrium will be assumed anywhere in the system, within each homogeneous phase and across the phase boundaries between the membrane and the solution. The phase rule, F = n − P + 1 (at constant temperature) may be applied to a region extending across a phase boundary, so that the number of independent intensive variables or degrees of freedom can be determined. The *number of components*, n, is equal to 4, and the *number of phases*, P, is equal to 2. This gives the *number of degrees of freedom*, F, equal to 3. One can, for example, change the chemical potentials μ_1, μ_2 and μ_3 independently at both sides of the membrane, while μ_4 (μ_{HM}) and the pressure are dependent variables. Thus the potential differences $\Delta\mu_1$, $\Delta\mu_2$ and $\Delta\mu_3$ over the membrane are also independent variables.

We shall now consider the stationary state for the system. At any position inside the membrane, the gradients in chemical potential are determined by the composition and the pressure of the electrolytes on the two sides of the membrane. The gradient $\nabla\mu_1$ is mainly a function of the difference $\Delta\mu_1$ over the membrane, but modified by $\Delta\mu_2$ and $\Delta\mu_3$:

$$\nabla\mu_1 = f_1(\underline{\Delta\mu_1}, \Delta\mu_2, \Delta\mu_3) \tag{5.61}$$

Thus $\nabla\mu_1$ at a given position in the membrane can be changed by changing $\Delta\mu_1$, while $\nabla\mu_2$ and $\nabla\mu_3$ are kept constant by slight adjustments of $\Delta\mu_2$ and $\Delta\mu_3$. Similarly $\nabla\mu_2$ and $\nabla\mu_3$ are mainly functions of $\Delta\mu_2$ and $\Delta\mu_3$ respectively, and each can be changed independently of the others.

We have discussed that $t_1 + t_2 = 1$ and $J_1 + J_2 = j$. With these equations and eqs. (5.59a,b) we obtain the following:

$$(l_{11} + l_{22})\nabla\mu_1 + (l_{12} + l_{22})\nabla\mu_2 + (l_{13} + l_{23})\nabla\mu_3 = 0 \tag{5.62}$$

This equation is valid for any values of $\nabla\mu_1$, $\nabla\mu_2$ and $\nabla\mu_3$. Since these are independent variables, it follows that $l_{11} + l_{21} = 0$, $l_{12} + l_{22} = 0$ and $l_{13} + l_{23} = 0$, or written in a more general form,

$$l_{1i} + l_{2i} = 0; \quad (i = 1,2,3) \tag{5.63}$$

Let us return to eqs. (5.54a–d). All the four forces in the four equations are independent. By means of an outer voltage source $\Delta\varphi$ can be changed and thus $\nabla\varphi$ is independent of the chemical potential gradients. Since $J_1 + J_2 = j$, eqs. (5.54a–b) add up to give eq. (5.54d). Thus eq. (5.56) is only a single case of the following more general equation:

$$L_{1i} + L_{2i} = L_{4i}; \quad (i = 1,2,3,4) \tag{5.64}$$

Equation (5.64) is valid when there are three independent mass fluxes across a cation exchange membrane and electric charge is transported by two monovalent cations between electrodes reversible to a common anion. Equation (5.63), however, is independent of the kind of electrodes used. Compare eq. (5.64) to eqs. (4.58) and (4.59) showing relations between coefficients in a concentration cell containing aqueous solutions of HCl and NaCl.

In order to study the independence of forces, the case of stationary state was considered. The validity of eqs. (5.63) and (5.64), however, extends to systems not in stationary state, since the coefficients l_{ij} and L_{ij} are gradient independent. They are functions of local composition, p and T.

When assuming $L_{12} = 0$ and $\nabla\mu_3 = 0$, eqs. (5.54a) and (5.54b) become identical with the Nernst–Planck flux equations:

$$J_1 = -\, L_{14}(\nabla\mu_1 + \nabla\varphi) \tag{5.65a}$$

$$J_2 = -\, L_{24}(\nabla\mu_2 + \nabla\varphi) \tag{5.65b}$$

(cf. Sections 4.1.2 and 4.1.4.).

Equations (5.54) contain 16 phenomenological coefficients. The Onsager reciprocal relations give six independent equations relating coefficients. Equation (5.64) gives four more independent equations relating coefficients. Thus a minimum of six phenomenological coefficients need to be found experimentally in order to determine all the 16 coefficients.

With $j = 0$, eqs. (5.59) contain nine diffusion coefficients. The Onsager relations give three independent equations relating coefficients. Equation (5.63) gives three more independent equations relating coefficients. Thus a minimum of three diffusion coefficients need to be found experimentally in order to determine all nine coefficients.

Generalization of the derivation. The above derivation can be extended to an n-component system containing ions with different charges.[3]

For a system of n components one may choose one of the components as the frame of reference, and thus the flux of this component is equal to zero. For an isothermal two-phase system without an electric current, there are n − 1 *independent* forces. With an electric current passing through the system, there are n *independent* forces (including the chemical potential gradients and the electric potential gradient) and n fluxes (including the mass fluxes $J_1, \ldots . J_{n-1}$ and the electric current density $J_n = j$). Fluxes may be linearly dependent:

$$\sum_{i=1}^{n} \alpha_i J_i = 0 \tag{5.66}$$

where α_i is the *coupling coefficient*. The set of coupling coefficients expresses the interdependence of the fluxes. For the system just dealt with ($HCl - NaCl - H_2O - HM$), $\alpha_1 = 1$, $\alpha_2 = 1$, $\alpha_3 = 0$ and $\alpha_4 = -1$. For a system $HCl - CaCl_2 - H_2O - HM$ then $\alpha_1 = 1$, $\alpha_2 = 2$, $\alpha_3 = 0$ and $\alpha_4 = -1$ when using a cation exchange membrane and anion-reversible electrodes.

With linearly related fluxes, the phenomenological coefficients are also linearly related:

$$\sum_{i=1}^{n} \alpha_i L_{ij} = 0; \quad (j = 1, \ldots n) \tag{5.67}$$

Similarly, the diffusion coefficients are linearly related:

$$\sum_{i=1}^{n-1} \alpha_i l_{ij} = 0 \; ; \quad (j = 1, \ldots n-1) \tag{5.68}$$

The values of the coupling coefficients are determined by the selectivity of the electrodes, the selectivity of the membrane (e.g. cation or anion exchange membrane) and the charges of the ions migrating through the membrane.

A set of coupling coefficients can be used to express any kind of interdependence between fluxes. Take, for example, the system HCl – NaCl – H_2O – HM. If all transport of water molecules is connected to the transport of components 1 and 2, this can be expressed by the following equation:

$$\alpha_1 J_1 + \alpha_2 J_2 = J_3 \tag{5.69}$$

where α_1 and α_2 are the numbers of water molecules transported together with components 1 and 2 respectively. This will give the following equation relating coefficients connected to transport of water:

$$\alpha_1 L_{1i} + \alpha_2 L_{2i} = L_{3i}; \quad (i = 1,2,3,4) \tag{5.70}$$

In a set of equations describing n fluxes caused by n forces, there are n^2 coefficients. According to the Onsager relations the number of independent coefficients is only $\frac{1}{2}n(n+1)$. For a system described by eq. (5.66), the n relations given in eq. (5.67) reduce the number of independent coefficients to $\frac{1}{2}n(n-1)$. This means that $\frac{1}{2}n(n-1)$ independent transport coefficients must be determined experimentally in order to obtain a complete description of isothermal transport of mass and charge across a cation exchange membrane between anion-reversible electrodes. When cation-reversible electrodes are used, one will arrive at the same number of independent coefficients.

5.5.3 *Experimental determination of phenomenological coefficients*

The relations between coefficients reduce the number of experimental determinations necessary for a complete description of the irreversible process. Furthermore, the relations give choices between different types of experiments, and the more easily available ones may be chosen. In some cases one may want to carry out entirely different experiments to determine the same coefficient, thus checking the reliability of the experimental data.

Returning to the previously considered example, the system HCl – NaCl – H_2O – HM with a cation exchange membrane and anion-reversible electrodes, we shall discuss the experiments needed to determine six independent coefficients. The experiments are arranged to have the vector fluxes perpendicular to the parallel surfaces of the membrane. This allows the gradients to be replaced by one-dimensional differentials in the set of flux equations (5.54). Further, the differentials are replaced by differences over a unit length of membrane thickness and fluxes over the whole cross-section are considered. Measurements are carried out at stationary state and differences are small. This allows the use of average values for the coefficients over the composition interval considered (see Section 3.5).

The coefficients L_{14}, L_{24}, L_{34} and L_{44}. The electric conductivity of the membrane, L_{44}, can be obtained from conductivity measurements. A Hittorf experiment gives either t_1 or t_2. When L_{44} is known, L_{14} and L_{24} are obtained applying eqs. (5.55a), (5.55b) and (5.56).

The coefficient L_{34} can be obtained by three different methods:

(1) The coefficient L_{34} can be obtained by measuring the flux of water, J_3, when a known current passes through the system with identical electrolytes and the same pressure on both sides of the membrane. When L_{44} is known, L_{34} is obtained applying eq. (5.55c).

(2) Under the same conditions as above, one can measure the electro-osmotic flux, i.e. the volume flux across the cell divided by the electric current. One obtains (cf. eq. (5.24)) the following equation:

$$t_3V_3 = (J_V/I)_{\Delta p,\Delta\mu_i=0} - (t_1V_1 + t_2V_2) - \Delta V_{el} \tag{5.71}$$

When L_{44} and t_1 (and t_2) are known, one obtains the transference coefficient for water and, by applying eq. (5.55c), one obtains L_{34}.

(3) One can obtain the coefficeint L_{34} by measuring the streaming potential, i.e. the electric potential difference across the cell divided by the pressure difference at zero current with identical electrolytes on both sides. One obtains (cf. eq. (5.23)) the following equation:

$$t_3V_3 = -\left[\frac{\Delta\varphi_{obs}}{\Delta p}\right]_{I,\Delta\mu_i(c)=0} - (t_1V_1 + t_2V_2) - \Delta V_{el} \tag{5.72}$$

One finds L_{34} from known quantities in the same way as for method (2).

The coefficients L_{11}, L_{12} *and* L_{22}. These coefficients can all be obtained from a pure diffusion experiment when L_{14}, L_{24} and L_{44} are known. When I is equal to zero, eq. (5.59a) is reduced to $J_1 = -\bar{l}_{11}\Delta\mu_1 - \bar{l}_{12}\Delta\mu_2 - \bar{l}_{13}\Delta\mu_3$, which, combined with eq. (5.63), gives

$$J_1 = -\bar{l}_{11}(\Delta\mu_1 - \Delta\mu_2) - \bar{l}_{13}\Delta\mu_3 \tag{5.73}$$

The experiment can be arranged with $\Delta\mu_3 = 0$ (the vapour pressure over both electrolytes is the same), or one can correct for the small term $\bar{l}_{13}\Delta\mu_3$ in the measured values for J_1 at known differences $\Delta\mu_1 - \Delta\mu_2$. Using eq. (5.60) and the known values of L_{14}, L_{24} and L_{44}, one obtains the coefficients L_{11}, L_{12} and L_{22}.

The coefficients L_{13} *and* L_{23}. In the pure diffusion experiment based on eq. (5.72) described above, one can measure J_3 in addition to J_1. From eqs. (5.59a), (5.59c) and (5.63) one obtains the ratio between the fluxes:

$$(J_3/J_1)_{I,\Delta\mu_3=0} = \bar{l}_{31}/\bar{l}_{11} \tag{5.74}$$

Thus $\bar{l}_{31}$ can be found after determining $\bar{l}_{11}$. When L_{14}, L_{24} and L_{34} are known, L_{13} and L_{23} are obtained by applying eqs. (5.60) and (5.64).

The coefficient L_{33}. This coefficient can be obtained by measuring the flux of water through the membrane at zero current with identical electrolytes, but different pressures on the two sides of the membrane. Then $\Delta\mu_i = V_i\Delta p$. One can use these expressions in eq. (5.59c) and at the same time eliminate $\bar{l}_{32}$ by means of eq. (5.63). Then the flux of water is

$$J_3 = -\bar{l}_{31}(V_1 - V_2)\Delta p - \bar{l}_{33}V_3\Delta p$$

A rearrangement of the above equation gives

$$\bar{l}_{33} = -1/V_3\,(J_3/\Delta p) - \bar{l}_{13}\,(V_1 - V_2)/V_3 \tag{5.75}$$

When L_{13}, L_{14}, L_{34} and L_{44} are known, L_{33} is obtained applying eq. (5.60).

5.6 *The Contribution to the Dissipation Function from Laminar Flow*

We shall consider systems with a *laminar flow* in addition to diffusion, but with no macroscopic kinetic energy change. In order to simplify the treatment, we shall consider forces and fluxes in the x-direction only. We shall limit the treatment to situations of laminar (viscous) flow with no turbulence and no slip of the liquid at the wall. Figure 5.9 illustrates laminar flow through a tube. The liquid in contact with the wall forms an infinitely thin stationary layer, and the velocity of the liquid increases gradually towards the centre of the tube. A pressure difference between the ends of the tube forces the liquid through the tube against the *viscous drag* at the wall. The mechanical work is converted to heat by this process, i.e. it is dissipated.

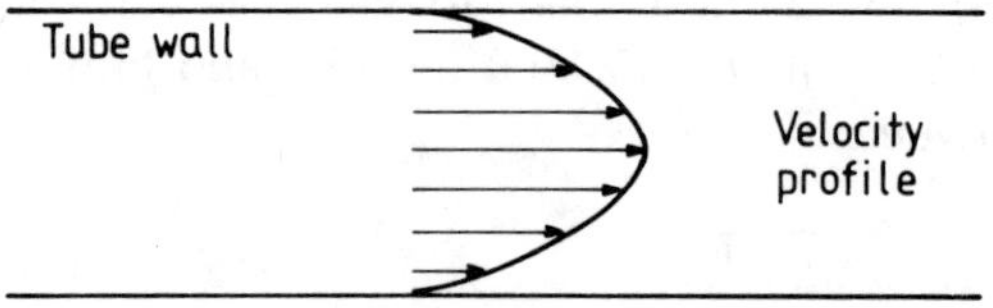

Figure 5.9. Laminar flow through a tube.

The membrane may be considered as a system of small pores and channels, where there is laminar flow in addition to the diffusion. The flux of a component anywhere in the membrane is thus composed of a flux with respect to the centre of gravity in a volume element and a flux of the volume element itself under the force of the pressure gradient. The dissipation function for the whole process, $T\Theta$, is the sum of the dissipation function for the diffusion with the *barycentre* (centre of gravity) as the frame of reference, $(T\Theta)_{\text{dif}}$, and the dissipation function for the flux of the volume element itself, $(T\Theta)_{\text{lam}}$:

$$T\Theta = (T\Theta)_{\text{dif}} + (T\Theta)_{\text{lam}} \tag{5.76}$$

In stationary state the barycentre has a constant velocity, v, with respect to the immobile frame of reference, the membrane, and the velocity of component i with respect to the same frame of reference is v_i. The velocity of the barycentre is the weighted sum of the velocities of the components:

$$v = \sum_{i=1}^{n} \frac{M_i c_i v_i}{\rho} \tag{5.77}$$

where M_i is the *molar mass of component* i and ρ is the *density of the mixture*. The weight fraction of component i in the volume element is $M_i c_i/\rho$. The flux of component i *with respect to the barycentre*, J_i^B, is

$$J_i^B = c_i(v_i - v) \tag{5.78}$$

where $(v_i - v)$ is the velocity of component i with respect to the barycentre. The conjugate force is given by the gradient in chemical potential. With a gradient in pressure, the gradient in chemical potential is given by eq. (5.16). In the x-direction we have

$$X_i = -\,d\mu_i/dx = -\,d\mu_i(c)/dx - V_i dp/dx \tag{5.79}$$

The dissipation function for the diffusion is the sum of the force – flux products:

$$\begin{aligned}(T\Theta)_{dif} &= \Sigma_1^n X_i J_i^B = \\ &-\Sigma(c_i d\mu_i(c)/dx)\,(v_i - v) - (dp/dx)\Sigma c_i V_i (v_i - v)\end{aligned} \tag{5.80}$$

The dissipation function for the laminar flow is

$$(T\Theta)_{lam} = -\,(dp/dx)v \tag{5.81}$$

Introducing the Gibbs–Duhem equation, $\Sigma c_i d\mu_i(c)/dx = 0$, and the relation between the partial molar volumes, $\Sigma c_i V_i = 1$, eq. (5.80) is reduced to

$$(T\Theta)_{dif} = -\,\Sigma(c_i d\mu_i(c)/dx)v_i - (dp/dx)\Sigma c_i V_i v_i + (dp/dx)v \tag{5.82}$$

and the dissipation function for the whole process (eq. (5.76)) can be written

$$T\Theta = -\,\Sigma(c_i d\mu_i(c)/dx)v_i - (dp/dx)\Sigma c_i V_i v_i \tag{5.83}$$

It can be seen that eq. (5.83) also represents the sum of the force–flux products when the immobile *membrane is the frame of reference* for the fluxes of the components, J_i^M:

$$J_i^M = c_i \boldsymbol{v}_i \tag{5.84}$$

while the forces are still given by eq. (5.79). The membrane is component n, and $J_n^M = 0$. Thus the dissipated energy for the whole process can be written

$$T\Theta = \Sigma_1^{n-1} X_i J_i^M \tag{5.85}$$

When there is no pressure difference ($\nabla p = 0$), there is no transport of bulk liquid and $(T\Theta)_{lam} = 0$, and we may choose either frame of reference. The Onsager reciprocal relations are then valid regardless of the frame of reference. When there is a transport of bulk liquid, the Onsager relations are valid for coefficients in flux equations for J_i^B, but generally not for coefficients in the flux equations for J_i^M. For more details see Prigogine,[11] de Groot and Mazur,[3] Mason and del Castillo,[12] and Lorimer.[13]

It is very inconvenient to use a moving barycentre as the frame of reference. In an experimental situation an immobile frame of reference is needed. When fluxes are measured with the membrane as a frame of reference, each flux is composed of a diffusional part, J_i^B, and a convective part due to the laminar flow, $c_i\boldsymbol{v}$:

$$J_i^M = J_i^B + c_i \boldsymbol{v} \tag{5.86}$$

where J_i^B is defined by eq. (5.78). It will be assumed that in a volume element the diffusion with the barycentre as frame of reference is independent of the flux of the volume element itself, i.e. that $(\boldsymbol{v}_i - \boldsymbol{v})$ is independent of the value of $\boldsymbol{v}$. Then the coefficients in the following flux equation:

$$J_i^B = \sum_{j=1}^{n-1} l_{ij} X_j \tag{5.87}$$

are independent of the laminar flow.

Experimentally determined fluxes in a membrane involve averages over the membrane cross-section. It will be assumed that eqs. (5.85) and (5.86) are also valid for these averages. By laminar flow the velocity of the liquid varies with the distance from a solid surface (see Fig. 5.9). The velocity v is then an average of the different velocities through the y,z-plane. In the stationary state it is a linear function of the pressure difference:

$$v = - L_o \Delta p \tag{5.88}$$

where L_o is a parameter inversely proportional to the viscosity coefficient of the liquid, and depending on the structure of the membrane. In the stationary state v, J_i^B and J_i^M are constant through the membrane (in the x-direction). In practical measurements a difference in chemical potential and a pressure difference are applied over the membrane and the coefficients obtained are average values over the membrane (in the x-direction) (cf. Section 3.5).

Experimental values for J_i^B and the coefficients $\overline{l}_{ij}$ are obtained by measurements when $\Delta\mu_i \neq 0$ and $\Delta p = 0$ (see Section 5.5). Experimental values for $c_i v$ are obtained by measurements when $\Delta\mu_i = 0$ and $\Delta p \neq 0$.

5.7 *References*

1. Harned, H. S. and Owen, B. B., *The Physical Chemistry of Electrolytic Solutions*, Reinhold, New York, 3rd ed., 1958.
2. Lobo, V. M. M., *Electrolyte Solutions, Literature Data on Thermodynamic and Transport Properties*, Universitade de Coimbra, Coimbra, Vol. 1 1975 (reprinted 1984), Vol. II, 1981 (with coauthor Quaresma, J. L.)
3. de Groot, S. R. and Mazur, P., *Non-Equilibrium Thermodynamics*, North-Holland, Amsterdam, 1962.
4. van't Hoff, J. H., *Z. Phys. Chem.*, **1**, 481 (1887).
5. Donnan, F. G., *Z. Elektrochem.*, **17**, 572 (1911).
6. Stavermann, A. J., *Rec. trav. chim.*, **70**, 344 (1951).
7. Stavermann, A. J., *Trans. Faraday Soc.*, **48** 176 (1952).
8. Katchalsky, A. and Kedem, O., *Biophys. J.*, **2** (suppl.), 53 (1962).
9. Katchalsky, A. and Curran, P. F., *Nonequilibrium Thermodynamics in Biophysics*, Harvard University Press, 1965.
10. Førland, K. S., Førland, T. and Ratkje, S. K., *Acta Chem. Scand.*, **A31**, 47 (1977).
11. Prigogine, I., *Etude Thermodynamique des Phénomènes Irréversibles*, Desoer, 1947.
12. Mason, E. A. and del Castillo, L. F., *J. Membrane Sci.*, **23** 199 (1985).
13. Lorimer, J. W., *J. Membrane Sci.*, **14** 275 (1983).

5.8 *Exercises*

5.1. *Equilibrium membrane – electrolyte interface, NaM + HCl ⇄ HM + NaCl.* Consider the exchange of Na^+ and H^+ between a cation exchange membrane and a dilute electrolyte solution of chlorides. The following corresponding values for mole fraction HM in membrane, x_{HM}, and mole fraction HCl in solution, $x_{HCl} = HCl/(n_{HCl} + n_{NaCl})$, were observed at 25°C:

x_{HM} = 0.265 0.490 0.732 0.846

x_{HCl} = 0.200 0.400 0.600 0.800

Assume ideal solution both in the membrane and in the electrolyte and find an approximate value for the equilibrium constant.

5.2. *The emf of concentration cells with membrane.* The half-cells are separated by a perfect cation exchange membrane, $|^C|$, in both cells (a) and (b) below. Assume ideal solutions and calculate the emf of each cell at 25°C, when c_I = 0.1 kmol m^{-3} and c_{II} = 0.01 kmol m^{-3}. Neglect the contribution to the emf from transference of water.

(a) Ag(s)|AgCl(s)|HCl(aq,c_I)$|^C|$HCl(aq,c_{II})|AgCl(s)|Ag(s)

(b) (Pt)H_2(g,1 atm) |HCl(aq,c_I)$|^C|$HCl(aq,c_{II})| H_2(g,1 atm)(Pt)

5.3. *Contribution to the emf from the transference of water.* Consider the cell, Ag(s)|AgCl(s)|HCl(aq,c_I)$|^A|$HCl(aq,c_{II})|AgCl(s)|Ag(s). The anion exchange membrane, $|^A|$, is not perfect, $t_{H^+} = 0.05$. Water is transferred mainly with Cl^-, and $t_{H_2O} = -5$. Assume ideal solutions and constant transference numbers. Temperature is 25°C. Express the emf of the cell as a function of c_I and c_{II}. Give the contributions from the transfer of H_2O and of H^+ as separate terms, $\Delta\varphi_1$ and $\Delta\varphi_2$ respectively. Let c_I be 0.01 kmol m^{-3}, and calculate numerical values for $\Delta\varphi_1$ and $\Delta\varphi_2$ when c_{II} is (i) 0.01 kmol m^{-3}, (ii) 0.05 kmol m^{-3}, (iii) 0.1 kmol m^{-3}.

5.4. *Influence of pressure difference.* The following cell is considered: Ag(s)|AgCl(s)| HCl(aq,c_I),$p_I|^C|$ HCl(aq,c_{II}), p_{II}|AgCl(s)|Ag(s). The pressure difference across the cation exchange membrane, $|^C|$, is $\Delta p = p_{II} - p_I$. The emf observed for the cell is 120 μV when Δp = 300 mm Hg. Use the values for partial molar volumes, V_{Ag} = 10.3 $cm^3\ mol^{-1}$ and V_{AgCl} = 25.8 $cm^3\ mol^{-1}$, and calculate the electric force, $-\Delta\varphi$, for transport in the cell.

5.5. *Streaming potential.* The streaming potential is studied experimentally using the cell Ag(s)|AgCl(s)|HCl(aq,c)$p_I|^C|$HCl(aq,c)p_{II}| AgCl(s)|Ag(s). The membrane, $|^C|$, conducts by H^+ only. The emf of the cell, E, was observed for different values of $\Delta p = p_{II} - p_I$

when $c = 0.1$ kmol m^{-3}:

Δp/cm H_2O	= 40	80	160	200
$-E/\mu V$	= 1.3	2.5	5.1	6.5

Calculate the transference coefficient for water, t_{H_2O}, using the following values for partial molar volumes, all with the dimension cm^3 mol^{-1}:

$V_{HCl} = 18.4$, $V_{H_2O} = 18.0$, $V_{Ag} = 10.3$, $V_{AgCl} = 25.8$

5.6. *The emf of a cell containing two salts in the electrolyte.* Consider the cell

$Ag(s)|AgCl(s)|HCl(aq,c_I)|^C|HCl(aq,c_{II}),\ CaCl_2\ (aq,c_{III})|AgCl(s)|Ag(s)$.

The membrane, $|^C|$, is cation conducting, and the concentration of chloride ions is the same in both electrolytes, $c_{Cl^-} = c_I = c_{II} + 2\,c_{III}$. Neglect the very small contribution to the emf of the cell from transfer of H_2O, and show that the emf then can be expressed by the equation

$$E = -\frac{RT}{F}\int_{c_I}^{c_{II}}\left(\frac{t_{H^+}}{c_{H^+}} - \frac{\frac{1}{2}\,t_{Ca^{2+}}}{c_{Cl^-} - c_{H^+}}\right)dc_{H^+}$$

where concentrations refer to an ideal electrolyte solution of constant chloride ion concentration in equilibrium with the membrane.

5.7. *Interdiffusion: relations between coefficients.* Two aqueous solutions of KCl and $SrCl_2$ are separated by a cation exchange membrane. (a) Write down the set of flux equations for the system. (b) Which relations are there between the coefficients? How many independent coefficients are there in the flux equations? (c) Assume $\nabla\mu_{H_2O} = 0$ throughout the system and express $J_2 = J_{Sr^{2+}}$ by means of one coefficient, l_{22}, and the gradients in chemical potential for KCl and $SrCl_2$. (d) A diffusion experiment was carried out under the following conditions: thickness of membrane, 0.5 mm; temperature, 25°C; $c_{Cl^-} = 0.03$ kmol m^{-3} in both solutions; on the left-hand side the equivalent fraction, $x_{Sr^{2+},\ell} = 0.5$; on the right-hand side the equivalent fraction, $x_{Sr^{2+},r} = 0.4$. The flux $J_{Sr^{2+}}$, under these conditions, was equal to 1.6×10^{-6} mol m^{-2} s^{-1}. Calculate an average value of the coefficient l_{22}. Assume ideal solutions of constant c_{Cl^-} in equilibrium with the membrane.

CHAPTER 6

Systems with Temperature Gradients

Gradients in natural logarithm of temperature, in chemical potentials, and in electric potential were the forces considered in Chapter 2. In a non-isothermal system heat can be transported by pure heat conduction without any flux of mass (see Section 2.1.1). The simultaneous transfer of heat and matter was discussed in Section 2.1.2. When there is a flux of mass, the *total heat flux*, J_{Φ}, contains the flux of components multiplied by their respective partial molar enthalpies, $\Sigma_i H_i J_i$, in addition to a *measurable heat flux*, J_q (cf. eq. (2.16)):

$$J_{\Phi} = J_q + \Sigma_i H_i J_i \tag{6.1}$$

The dissipation function for an irreversible process is expressed as the product of forces and fluxes (cf. eq. (2.29):

$$T\Theta = -\nabla \ln T J_q - \Sigma_i \nabla \mu_{i,T} J_i - \nabla \varphi j = \Sigma_j X_j J_j \tag{6.2}$$

The corresponding flux equations are

$$J_j = -L_{j1} \nabla \ln T - \sum_{i=2}^{k-1} L_{ji} \nabla \mu_{i,T} - L_{jk} \nabla \varphi \tag{6.3}$$

In an isothermal system the force $-\nabla \ln T = 0$, but there may still be a flux J_q by interaction with the other forces via the cross coefficients. This heat flux was not considered in Chapters 4 and 5, since it does not contribute to the dissipation function when $-\nabla \ln T = 0$.

The interdependence of heat and mass fluxes is the cause of some observable effects. The flux of heat caused by a concentration gradient is called the *Dufour effect*. The reciprocal effect, *thermal diffusion*, is a flux of matter caused by a temperature gradient. In liquids it is called the *Soret effect*. Thermal diffusion can be utilized for the separation of mixtures of isotopes, mixtures of close-boiling substances and for other difficult separations where the more conventional separation methods do not perform satisfactorily. The phenomenon of *thermal osmosis* can be observed across a membrane. A temperature difference across the membrane, leading to a flux of matter, can be counterbalanced by a pressure difference to stop the flux of matter. Closely connected to thermal osmosis is the phenomenon of frost heave. This will be discussed in Section 11.1.

The flux of heat caused by a gradient in electric potential is called the *Peltier effect*, while the flux of electric charge caused by a temperature gradient is called the *thermoelectric effect* or the *Seebeck effect*. The use of the thermocouple for temperature measurement is based on the Seebeck effect. Thermoelectric energy conversion, the interconversion of heat and electric energy for power generation or heat pumping, is based on Seebeck and Peltier effects. The practical applications with the use of semiconductors are numerous. Thermoelectric generators have low conversion efficiency, but are used for special applications where small size, reliability, easy maintenance, absence of moving parts and vibrations, lack of pollution problems, etc., are important. Thermoelectric heat pumps have the same drawbacks and advantages, and are useful for specialized applications.[1]

In electrochemical cells there may be heat, mass and charge fluxes combined. The additivity of the entropy production from the different force – flux products makes it possible to treat the effects separately. The heating or cooling of electrodes by the electrolytic process may be a serious problem. This will be discussed in Section 11.2.

The study of systems with temperature gradients involves some new problems compared to isothermal transport. The heat flux, J_q, and the corresponding entropy flux, $J_S = J_q/T$, are not constant over distance — not even under stationary conditions where J_i and j are constant. Furthermore we need knowledge about how both $\nabla\mu_{i,T}$ and $\nabla\varphi$ vary with the temperature in order to calculate the total entropy production in a system.

It is common to make some approximations to facilitate the calculation of transport parameters. When the stationary state is considered and differences are small, eq. (6.2) is replaced by eq. (3.22), and the flux eqs. (6.3) are replaced by eqs. (3.23). This implies that the variation in J_q over the distance considered is negligibly small compared to the value of J_q. In the mathematical derivations from these equations it is assumed that $\Delta \ln T \approx (1/T)\Delta T$ and $\Delta \ln c \approx (1/c)\Delta c$. This implies that the temperature difference, ΔT, is much smaller than T, and that Δc is much smaller than c.

6.1 *The Soret Effect and the Dufour Effect*

We shall consider a system consisting of a solvent, component 1, and a solute, component 2, in a temperature gradient as shown in Fig. 6.1. The container of the solution is of unit cross-section and unit length.

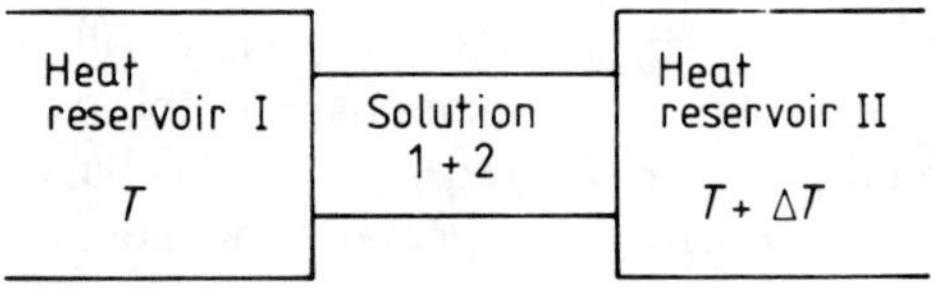

Figure 6.1. A solution of unit cross-section and unit length in a temperature gradient.

The temperature difference between the two sides of the system is maintained by means of heat reservoirs at the two temperatures T and $T + \Delta T$. There is a continual flux of heat from the warmer heat reservoir to the colder one. A flux of matter will build up concentration gradients in a solution originally having an even distribution of solute. The build-up continues until a stationary state is attained. Then the temperature gradient is just counterbalanced by the gradient in chemical potential. In the stationary state the heat flux continues, but there is no longer a flux of matter. (In an experimental situation it may take weeks or months to obtain stationary state.[2])

The container walls bordering the reservoirs of constant temperature may be taken as an immobile frame of reference for all movements in the solution. Then one must operate with two mass fluxes, the flux of the solute and the flux of the solvent relative to the container walls. The expressions for the fluxes are rather complex.[3,4] When the solution is dilute, the movement of the solvent relative to the container wall can be neglected. As an approximation one can operate with only one mass flux, the flux of the solute relative to the solvent.

Using the solvent as the frame of reference, there are two fluxes in the system, the flux of heat, J_q, and the flux of the solute, component 2, J_2. These fluxes can be expressed by the following flux equations:

$$J_q = -\bar{l}_{11}\Delta \ln T - \bar{l}_{12}\Delta\mu_{2,T} \tag{6.4a}$$

$$J_2 = -\bar{l}_{21}\Delta \ln T - \bar{l}_{22}\Delta\mu_{2,T} \tag{6.4b}$$

Since there is no flux of charge, diffusion coefficients, $\bar{l}_{ij}$, are used. As an approximation, gradients are replaced by differences and coefficients have average values.

The coefficient $\bar{l}_{11}$ is related to the *thermal conductivity*, λ. For a constant chemical potential, $\Delta\mu_{2,T} = 0$, we have $J_q = -\bar{l}_{11}\Delta \ln T = -(\bar{l}_{11}/T)\Delta T = -\lambda\Delta T$, and thus the thermal conductivity

$$\lambda = \bar{l}_{11}/T \tag{6.5}$$

The coefficient $\bar{l}_{22}$ is related to the *isothermal* Fick's diffusion coefficient, D. For constant temperature, $J_2 = -\bar{l}_{22}\Delta\mu_{2,T}$. For an ideal non-electrolyte solution we have $\Delta\mu_{2,T} = RT\Delta \ln c_2 = RT\Delta c_2/c_2$ (cf. eq. (4.27b). This gives us $J_2 = -\bar{l}_{22}\ RT\Delta c_2/c_2$,which can be compared to Fick's first law of diffusion (eq. (4.28)). Here it can be written in the form $J_2 = -D\Delta c_2$, where D is the average diffusion coefficient over the concentration interval. We obtain the following equation for *Fick's diffusion coefficient*:

$$D = \bar{l}_{22}\ RT/c_2 \tag{6.6}$$

In dilute solutions $\bar{l}_{21}$ is found to be approximately proportional to c_2, and it is convenient to define a *thermal diffusion coefficient*, D_T,[5] as follows:

$$D_T = \bar{l}_{21}/(c_2 T) \tag{6.7}$$

Introducing Fick's diffusion coefficient (eq. (6.6)) and the thermal diffusion coefficient (eq. (6.7)), into the flux equation (6.4b) we obtain

$$J_2 = -D_T c_2 \Delta T - D \Delta c_2 \tag{6.8}$$

From eq. (6.8) it can be seen that there will be a flux of solute creating a difference in composition in a solution initially of constant composition, but with a temperature difference, the *Soret effect*. Conversely, the flux of solute caused by a concentration difference will create a temperature difference in the solution (unless heat is removed from the hot end of the diffusion zone and replenished at the cold end), the *Dufour effect*.

The ratio between the thermal diffusion coefficient and the Fick's diffusion coefficient is called the *Soret coefficient*, s_T. For the system in a stationary state, where $J_2 = 0$, the Soret coefficient can be expressed by the differences in concentration and temperature:

$$s_T = D_T/D = -\left[\frac{\Delta c_2}{c_2 \Delta T}\right]_{J_2=0} \tag{6.9}$$

Another quantity frequently used in the description of transfer phenomena in non-isothermal systems is the *heat of transfer* (or *heat of transport*), q_i^*. It is the heat transferred from left to right, coupled to the transfer of component i at zero temperature difference. In our case, with only one mass flux, we have from eqs. (6.4) the following:

$$q_2^* = (J_q/J_2)_{\Delta T=0} = \bar{l}_{12}/\bar{l}_{22} \tag{6.10}$$

The relationship between s_T and q_2^* can be seen when D and D_T in eq. (6.9) are replaced by eqs. (6.6) and (6.7). Introducing the Onsager reciprocal relation for average coefficients, $\bar{l}_{12} = \bar{l}_{21}$, we obtain the following for a dilute solution:

$$s_T = q_2^*/RT^2 \tag{6.11}$$

6.2 *Thermal Osmosis*

Transfer of mass across a phase boundary leads to absorption or release of heat when the partial molar enthalpy of the transferred component is not the same in the two phases. This may give rise to a strong interaction between the mass flux and the temperature gradient.

We shall consider a system consisting of a membrane and a liquid that dissolves in the membrane. A temperature difference is created across the membrane by means of heat reservoirs at the temperatures T and $T + \Delta T$. The temperature difference causes a flux of liquid through the membrane, which leads to a pressure difference between the two sides. This phenomenon is called *thermal osmosis*, and the system is illustrated in Fig. 6.2.

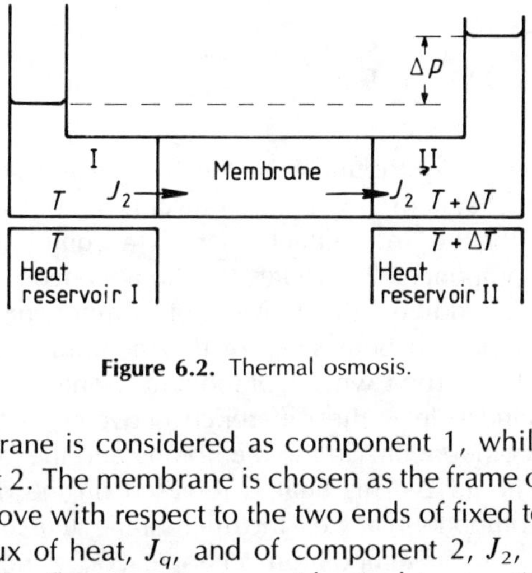

Figure 6.2. Thermal osmosis.

The membrane is considered as component 1, while the liquid is component 2. The membrane is chosen as the frame of reference. It does not move with respect to the two ends of fixed temperature. There is a flux of heat, J_q, and of component 2, J_2, through the membrane. The flux equations are identical to eqs. (6.4):

$$J_q = -\bar{l}_{11}\Delta \ln T - \bar{l}_{12}\Delta\mu_{2,T} \tag{6.4a}$$

$$J_2 = -\bar{l}_{21}\Delta \ln T - \bar{l}_{22}\Delta\mu_{2,T} \tag{6.4b}$$

Since there is pure component 2 on both sides of the membrane, the expression for $\Delta\mu_{2,T}$, however, differs from the one used for a solute in Section 6.1:

$$\Delta\mu_{2,T} = V_2\Delta p \tag{6.12}$$

Here V_2 is the molar volume of component 2.

When a stationary state is attained for fixed values of T and ΔT, we have $J_2 = 0$. Then an expression for the *thermal osmotic pressure*, Δp, is obtained from eqs. (6.4b) and (6.12):

$$\Delta p = -(1/V_2)(\bar{l}_{21}/\bar{l}_{22})\,\Delta \ln T \tag{6.13}$$

Since $\bar{l}_{21} = \bar{l}_{12}$, the ratio $\bar{l}_{21}/\bar{l}_{22}$ can be identified as the ratio between eqs. (6.4a) and (6.4b) at constant temperature:

$$(J_q/J_2)_{\Delta T=0} = \bar{l}_{12}/\bar{l}_{22} = q_2^* \tag{6.14}$$

(cf. eq. (6.10)). Thus we have

$$\Delta p = -(q_2^*/V_2)\Delta \ln T \tag{6.15}$$

There are two contributions to the heat of transfer, q_2^*, in the heterogeneous system involving transport across a membrane. Similarly as for the Soret effect, there is a contribution from the transfer of component 2 through the homogeneous phase of the membrane. In addition the transfer of component 2 across the phase boundaries on both sides of the membrane contributes to q_2^*. The heat absorbed when component 2 enters the membrane on the lefthand side is the difference between the partial molar enthalpy of component 2 in the membrane and the molar enthalpy of the pure liquid 2. This heat is released on the righthand side when the component leaves the membrane (the change in the enthalpies with the temperature change, ΔT, is neglected). The heat may be positive or negative. In many cases this contribution to q_2^* is the larger one.

In Section 6.1. the thermal conductivity was defined as $\lambda = \bar{l}_{11}/T$ (see eq. (6.5)). When there is no difference in chemical potential, we obtain from eq. (6.4a) $J_q = -(\bar{l}_{11}/T)\Delta T = -\lambda\Delta T$. In this situation there is a flux of component 2, and part of the heat is transferred with this component. It is possible to separate the heat transferred by pure conduction and the heat transferred with component 2. By eliminating $\Delta\mu_{2,T}$ from eqs. (6.4) and with eq. (6.14) we obtain

$$J_q = -\frac{\overline{l}_{11} - \overline{l}_{12}^2/\overline{l}_{22}}{T}\Delta T + q_2^* J_2 \tag{6.16}$$

In the stationary state, when $J_2 = 0$, all heat is transferred by pure conduction, and thus the expression $(\overline{l}_{11} - \overline{l}_{12}^2/\overline{l}_{22})/T$ is the *pure thermal conductivity* of the system.

6.3 *The Peltier Effect and the Seebeck Effect*

We shall consider a system consisting of two electronic conductors, A and B. They are of different composition and are joined at both ends to form a loop. The two junctions are kept at different temperatures, T and $T + \Delta T$, by means of heat reservoirs. A potentiometer is inserted in conductor A by means of electric leads. The junctions between A and the leads are both at the same temperature, T_o. The pressure is constant in the system. The system is illustrated in Fig. 6.3.

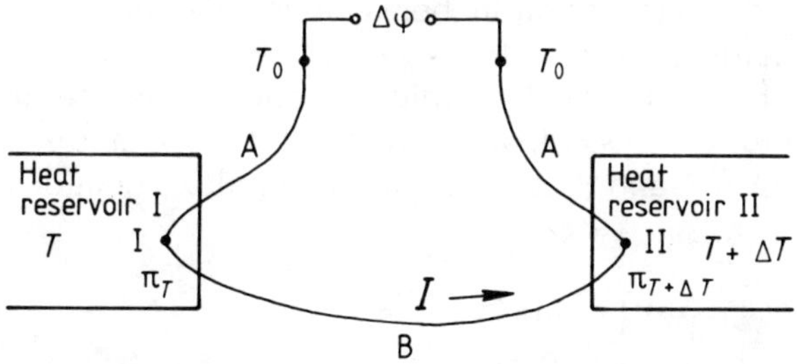

Figure 6.3. Arrangement for investigating the Peltier effect and the Seebeck effect (see text for explanation of the symbols).

The temperature difference between junctions I and II gives rise to an electric potential, the *Seebeck effect* or the *thermoelectric effect*. For small values of ΔT, the potential, $\Delta\varphi$, observed with the potentiometer at zero current, $I \approx 0$, is proportional to ΔT. The *thermoelectric power* is defined as the ratio $\Delta\varphi/\Delta T$.

When current is passed counter-clockwise through the loop, heat (positive or negative) is absorbed at junction I, while heat is released at junction II, the *Peltier effect*. The amount of heat is *proportional* to the current and depends on the temperature. It comes in addition to the Joule heat, which is proportional to the *square* of the current, RI^2, and the conducted heat, which is *independent* of the current. Since both junctions between A and the leads to the potentiometer are at the same temperature, T_o, the heat absorbed at one junction is counterbalanced by the heat released at the other. Thus the total change in this part of the loop is equal to zero.

We shall define the *Peltier heat* as the number of joules taken from the reservoir at the A–B junction by the transfer of 1 faraday positive charge in the direction from A to B. The Peltier heat is given the symbol π. (Peltier heat often refers to the transfer of 1 coulomb. The numerical value of π will then be smaller by a divisor of 96,500).

If both ΔT and π are positive and current is passed counter-clockwise through the loop, heat is transported from left to right, i.e. from low to high temperature. This means that electric work must be supplied from the potentiometer. Thus $\Delta\varphi$ is negative, in agreement with the sign convention, and the thermoelectric power, $\Delta\varphi/\Delta T$, is negative. We shall find mathematical expressions for the relation between the Peltier heat and the thermoelectric power.

Using the loop of electronic conductors as the frame of reference for our system, we have two fluxes, J_q and I, and two forces, $-\Delta\ln T$ and $-\Delta\varphi$. The following flux equations can be written:

$$J_q = -\bar{L}_{11}\Delta\ln T - \bar{L}_{12}\Delta\varphi \tag{6.17a}$$

$$I = -\bar{L}_{21}\Delta\ln T - \bar{L}_{22}\Delta\varphi \tag{6.17b}$$

At constant temperature T, the heat transferred from reservoir I to reservoir II per faraday, is the Peltier heat at the temperature T,[3]

$$(J_q/I)_{\Delta T=0} = \bar{L}_{12}/\bar{L}_{22} = \pi_T \tag{6.18}$$

The amount of heat transferred per unit time, J_q, is proportional to the current, I. By changing the sign of I, the direction is reversed for J_q. Since there is no temperature gradient, no heat is transported by heat conduction. For small values of I, the Joule heat is negligible and therefore the loss in energy by the transfer is negligible. The heat is transferred reversibly. The corresponding entropy transferred per faraday of charge is

$$S^* = \pi_T/T = \frac{1}{T}\bar{L}_{12}/\bar{L}_{22} \tag{6.19}$$

This equation shows that the entropy S^* is independent of gradient, which means that it is independent of the transport of heat by heat conduction when there is a temperature gradient.

Eq. (6.17b) is solved with respect to $\Delta\varphi$:

$$\Delta\varphi = -(\bar{L}_{21}/\bar{L}_{22})\Delta \ln T - (1/\bar{L}_{22})I \tag{6.20}$$

We shall neglect any change in S^* over a small temperature interval and use the approximation $\Delta \ln T = (1/T)\Delta T$. With the Onsager reciprocal relation $\bar{L}_{21} = \bar{L}_{12}$, the following is obtained from eqs. (6.19) and (6.20):

$$\Delta\varphi = -S^*\Delta T - I/\bar{L}_{22} \tag{6.21}$$

The change in S^* with temperature will be discussed in Section 6.4.

For a small current, $I \approx 0$, the ohmic resistance causes only a negligible potential loss, $I/\bar{L}_{22}$ is negligibly small and we have

$$\Delta\varphi = -S^*\Delta T = -(\pi_T/T)\Delta T \tag{6.22}$$

Thus the relation between the thermoelectric power and the Peltier effect is

$$\Delta\varphi/\Delta T = -S^* = -\pi_T/T \tag{6.23}$$

With the assumption of a constant S^* over the interval ΔT we have

$$S^* = \pi_T/T = \pi_{(T+\Delta T)}/(T + \Delta T) = -\Delta\varphi/\Delta T \tag{6.24}$$

or

$$\pi_{(T+\Delta T)} - \pi_T = S^*\Delta T = -\Delta\varphi \tag{6.25}$$

Equation (6.25) can be given a physical interpretation. When the values of S^* and ΔT are both positive, the heat reservoir II (at temperature $T + \Delta T$) receives more heat than is removed from heat reservoir I (at temperature T). The difference in heat per faraday, $S^*\Delta T$, is equal to the supplied electric work per faraday, $-\Delta\varphi$.

By reversing the direction of the current, the heat $\pi_{(T+\Delta T)}$ will leave heat reservoir II at temperature $T + \Delta T$, and the heat π_T will be absorbed by heat reservoir I at temperature T. The difference,. $- S^*\Delta T$, can be converted to electric energy.

6.4 *The Thomson Heat*

When the temperature differences in a system do not exceed some few degrees, eq. (6.23) gives a fairly accurate account of the relation between the thermoelectric power and the Peltier effect. The relation is based on the approximation that S^* does not change with temperature. For large differences in temperature, however, the Peltier heat cannot account fully for the electric energy conversion. We must also take into account the change in S^* with the temperature, which leads to a heat exchange with the surroundings when a current is passed through the system. This is the *Thomson heat*.

We shall consider a section of the A – B loop pictured in Fig. 6.3. The A – B junction at heat reservoir I and a part of the conductor B is pictured in Fig. 6.4. There is a temperature gradient over conductor B. The heat exchange with the surroundings for a small temperature interval, T_1 to T_2, is pictured as a heat exchange with a reservoir at the average temperature.

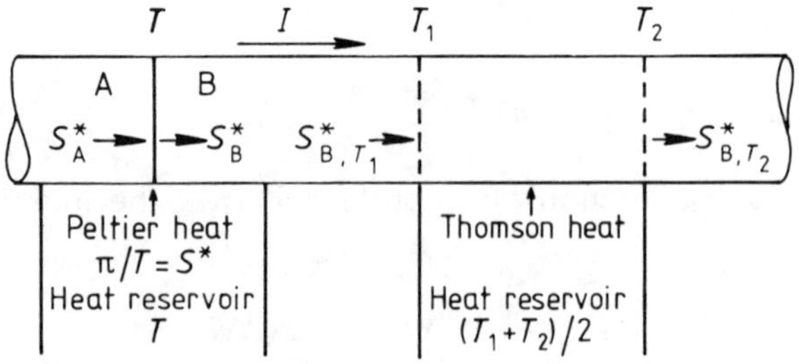

Figure 6.4. Arrangement for investigating Thomson heat (see text for explanation of the symbols).

First we shall consider the balance of entropy at the A – B junction where the Peltier heat is transferred reversibly from the heat reservoir at the temperature T. The corresponding entropy, S^*, is the difference between the entropy transported through the conductor away from the junction, S^*_B, and the entropy transported to the junction, S^*_A:

$$S^* = S^*_B - S^*_A \tag{6.26}$$

Both S^*_B and S^*_A, as well as S^*, refer to the transfer of 1 faraday of positive charge. In the B – A junction at the temperature $T + \Delta T$ (see Fig. 6.3) we have $- S^* = S^*_A - S^*_B$. The entropies S^*_A and S^*_B change with temperature and the variation in S^* with temperature can be attributed to these changes.

The change in S^*_B in a section of B from temperature T_1 to T_2 (see Fig. 6.4) represents a loss in entropy, $S^*_{B,T_2} - S^*_{B,T_1}$. In order to keep the temperature gradient undisturbed, this loss of entropy must be replenished from the surroundings, i.e. from the heat reservoir adjacent to the section. At constant pressure

$$S^*_{B,T_2} - S^*_{B,T_1} = (\partial S^*_B/\partial T)\Delta T \tag{6.27}$$

where $\Delta T = T_2 - T_1$ is small compared to T_1 and T_2. The heat absorbed from the heat reservoir, $T(\partial S^*_B/\partial T)\Delta T$, is called the *Thomson heat*. The heat absorbed over a temperature interval of 1 degree, $\Delta T = 1$, is called the *Thomson coefficient*, τ. For the conductor B we have

$$\tau_B = T(\partial S^*_B/\partial T)_p \cdot 1 = (\partial S^*_B/\partial \ln T)_p \tag{6.28a}$$

We find a similar expression for the Thomson coefficient in the conductor A:

$$\tau_A = T(\partial S^*_A/\partial T)_p \cdot 1 = (\partial S^*_A/\partial \ln T)_p \tag{6.28b}$$

The Thomson coefficient has the dimension $J\ K^{-1}$ faraday^{-1}, the same as a molar heat capacity or a molar entropy.

The Thomson heat is the heat supply needed to keep the entropy constant with time in the electric conductor; thus it is the heat absorbed reversibly by the conductor.

In the stationary state the temperature gradient is constant with time. When a current passes through the conductor, the resistance leads to a production of heat, RI^2. In order to keep a constant temperature gradient, this Joule heat less the reversible Thomson heat must be given off to the surroundings. The Thomson heat is proportional to I. When measuring the heat received by the surroundings for different values of I (positive and negative), one can obtain experimental values both for the Joule heat and for the Thomson heat. In the calculations one must also account for heat transferred through the conductor by heat conduction.

The Thomson coefficients τ_B and τ_A can be obtained from calorimetric measurements as functions of temperature. It has been shown experimentally that $\tau \rightarrow 0$ when $T \rightarrow 0$. If we postulate S_B^* and S_A^* equal to zero at 0 K, we can obtain absolute values for S_B^* and S_A^* at the temperature T by integrating eqs. (6.28a) and (6.28b):

$$S^*_{B,T} = \int_0^T \tau_B \mathrm{d}\ln T \tag{6.29a}$$

$$S^*_{A,T} = \int_0^T \tau_A \mathrm{d}\ln T \tag{6.29b}$$

These values of $S^*_{B,T}$ and $S^*_{A,T}$ can then be used to calculate the Peltier heat:

$$\pi_{AB,T} = T(S^*_{B,T} - S^*_{A,T}) \tag{6.30}$$

The calculated value of $\pi_{AB,T}$ can be compared to a value obtained from an independent experiment, to give experimental evidences of the theory.

6.5 *The Thermocouple*

The thermoelectric (Seebeck) effect is utilized for temperature measurements with the aid of a *thermocouple*. Most thermocouples consist of two metal wires, A and B, of different composition. The wires are welded together at one end, the *hot junction*. This junction is placed in or at the body of which the temperature is to be measured. The other end of each wire is placed in electric contact with a copper lead to the instrument for potential measurement. Both these contacts are kept at the same temperature, the *cold junction*. The difference in temperature between the hot and cold junctions gives rise to an electric potential which is measured at zero current, $I \approx 0$ (cf. eq. (6.22)). Figure 6.5 gives a schematic picture of a thermocouple.

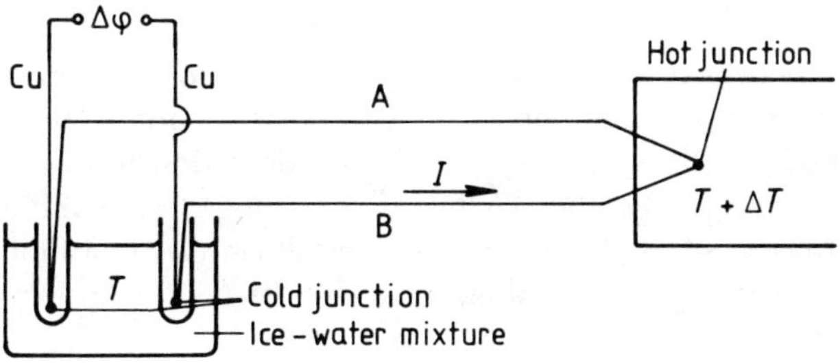

Figure 6.5. Coupling scheme for a thermocouple.

The thermocouple has three junctions between conductors of different composition — the B – A hot junction and the two cold junctions A – Cu and Cu – B. For each junction the transferred entropy per faraday, S^*, can be expressed by the difference between the entropy transported through the conductor away from the junction and up to the junction:

At the B – A hot junctions $\quad S^*_{BA} = S^*_A - S^*_B \qquad$ (6.31a)

At the A – Cu cold junction $\quad S^*_{ACu} = S^*_{Cu} - S^*_A$

At the Cu – B cold junction $\quad S^*_{CuB} = S^*_B - S^*_{Cu}$

At the same temperature, S^*_{Cu} has the same value at both cold junctions, and the total entropy transferred over the cold junctions is

$$S^*_{AB} = S^*_B - S^*_A \qquad (6.31b)$$

The nature of the conductor inserted between A and B does not make any difference when both cold junctions are kept at the same temperature. The same would be the case for the hot junction. This means that any gradual change in composition around the junction produced by the welding process will be without influence on the temperature measurement if the complete zone of varying composition is kept at the same temperature.

If we disregard the changes in S_A^* and S_B^* with temperature, we have $S_{AB}^* = -S_{BA}^*$; the entropy S_{AB}^* is transferred from the cold junction to the hot junction. Similarly to eq. (6.25) we have

$$\Delta\varphi = -S_{AB}^*\Delta T \tag{6.32}$$

Thermocouples are used for temperature measurements over large temperature differences; ΔT can be very large. The changes in S_A^* and S_B^* with temperature lead to a deviation from the linear relation between $\Delta\varphi$ and ΔT given by eq. (6.32). A calibration of the thermocouple is then needed to establish $\Delta\varphi$ as a function of ΔT. This is done by recording the value of $\Delta\varphi$ for known values of ΔT.

6.6 *Thermocells*

A *thermocell* is a non-isothermal electrochemical cell. The two electrodes are at different temperatures. One can observe a Seebeck effect; the temperature difference causes an electromotive force. A current through the cell causes Peltier effects.

In an electrochemical cell ions are current carriers in the electrolyte and there are chemical reactions at the electrodes, where electrons are taking over as current carriers through the electronic conductors. This is an important difference from systems of only electronic conductors, such as a thermocouple, where electrons are the only current carriers.

The electrolyte may be a solution composed of two or more components. Then concentration gradients and differences are possible, and we have a Soret effect in addition to the Seebeck effect and the Peltier effect. An effect similar to the Thomson heat in metals has not been reported for electrolytes.

6.6.1 *The cell (T)A|AX|A(T+ΔT), cell a*

The first cell that we shall consider consists of two electrodes of the metal A and a solid or liquid electrolyte AX. The lefthand-side electrode is kept at a constant temperature T by means of the heat reservoir I, while the right-hand-side electrode is kept at a constant temperature $T + \Delta T$ by means of the heat reservoir II. Both junctions between electrode and outer circuit are kept at the same temperature T_o. The cell is shown in Fig. 6.6.

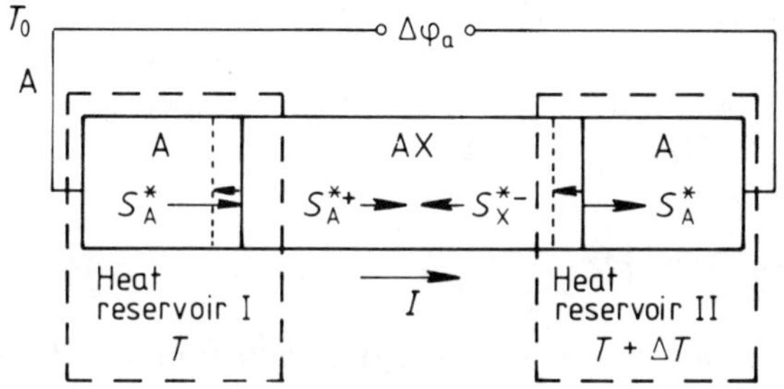

Figure 6.6. Schematic arrangement of the non-isothermal cell A|AX|A (see text for explanation of the symbols).

By a reasoning similar to that in Section 4.1.1, the following changes are found to be taking place upon the passage of one faraday of positive charges from left to right through the inner circuit of the cell. At the left-hand-side electrode one mole A is removed and t_{X^-} mole AX is supplied, while one mole A is supplied at the right-hand-side electrode and t_{X^-} mole AX is removed. When A is shifted from the left-hand-side electrode to the right-hand-side one in this way, the barycentre of the electrolyte, AX, moves relative to the walls of the container.

The container walls may be chosen as the frame of reference for the transports in the system. Since the barycentre of AX moves, we will have a flux J_{AX} in addition to J_q and I. Regardless of the frame of reference, the transfer of A is coupled to I by the simple relation $J_A = I$.

Constant pressure in the cell is assumed. Then $\Delta\mu_{AX,T} = 0$ for the pure salt AX and, similarly, $\Delta\mu_{A,T} = 0$ for the pure metal A. This means that neither J_{AX} nor J_A will contribute to the dissipated energy, since their conjugate forces are equal to zero.

There are only two forces in the system $-\Delta \ln T$ and $-\Delta\varphi_a$. By eq. (3.22) the dissipated energy per unit time is obtained:

$$\frac{TdS}{dt} = -\Delta \ln T J_q - \Delta\varphi_a I \tag{6.33}$$

The corresponding flux equations are:

$$J_q = -\bar{L}_{11}\Delta \ln T - \bar{L}_{12}\Delta\varphi_a \tag{6.34a}$$

$$I = -\bar{L}_{21}\Delta \ln T - \bar{L}_{22}\Delta\varphi_a \tag{6.34b}$$

The Peltier heat is defined as for systems of electronic conductors only (cf eq. (6.18)):

$$(J_q/I)_{\Delta T=0} = \bar{L}_{12}/\bar{L}_{22} = \pi_{T,a} \tag{6.35}$$

The electric potential for $I = 0$ is obtained from eq. (6.34b):

$$\Delta\varphi_a = -(\bar{L}_{21}/\bar{L}_{22})\Delta \ln T \tag{6.36}$$

By the Onsager reciprocal relations we have $\bar{L}_{12} = \bar{L}_{21}$, and for small temperature differences we can write $\Delta \ln T \approx (1/T)\Delta T$. Hence we have

$$\Delta\varphi_a = -\pi_{T,a}\Delta \ln T \approx -(\pi_{T,a}/T)\Delta T \tag{6.37}$$

where $\pi_{T,a}/T$ is the entropy transferred.

We shall consider the balance of entropy across the metal – electrolyte interface for the left-hand-side electrode when one faraday of positive charge passes from left to right through the cell (inner circuit). The interface receives entropy from the heat reservoir, $\pi_{T,a}/T$, entropy transported through the metal to the interface, S^*_A, and entropy transported through the electrolyte to the interface, $t_{X^-}S^*_{X^-}$. The disappearance of one mole A liberates the molar entropy, S_A. This balances the entropy consumed by the formation of t_{X^-} mole AX with a molar entropy, S_{AX}, and the entropy transported through the electrolyte away from the interface, $t_{A^+}S^*_{A^+}$;

$$\pi_{T,a}/T + S^*_A + S_A + t_{X^-}S^*_{X^-} = t_{X^-}S_{AX} + t_{A^+}S^*_{A^+} \qquad (6.38)$$

This is the reversible entropy balance. With a temperature difference there is a time-dependent irreversible entropy production in addition. That does not change the reversible entropy balance. Rearranging eq. (6.38) we obtain

$$\pi_{T,a}/T = - S_A - S^*_A + (t_{X^-}S_{AX} + t_{A^+}S^*_{A^+} - t_{X^-}S^*_{X^-}) \qquad (6.39)$$

The values of t_{A^+} and t_{X^-} depend on the chosen frame of reference, while we always have $t_{A^+} + t_{X^-} = 1$. The entropy transferred is independent of the chosen frame of reference, and the term inside the bracket in the equation above has the same value independent of frame of reference. From this we can infer that

$$S_{AX} = S^*_{A^+} + S^*_{X^-} \qquad (6.40)$$

From eqs. (6.37), (6.39) and (6.40) we obtain

$$\Delta\varphi_a/\Delta T = - \pi_{T,a}/T = (S^*_A + S^*_{X^-}) - (S_{AX} - S_A) = S_A + S^*_A - S^*_{A^+} \qquad (6.41)$$

The term $(S^*_A + S^*_{X^-})$ contains transport quantities. The entropy transported through the metal, S^*_A, is small, showing only small variations with the nature of the metal. For semiconductors one may find larger variations in the entropy transport. The entropy transported through the electrolyte, $S^*_{X^-}$, may contribute substantially to the Peltier heat. The term $(S_{AX} - S_A)$ contains thermodynamic entropies and originates from the chemical reaction. This term may give the major contribution to the Peltier heat. The Peltier heat of thermocells is usually much larger than the Peltier heat of thermocouples. There is no chemical reaction in a system of electronic conductors.

6.6.2 *The cell (T)X$_2$|AX|X$_2$(T+ΔT), cell b*

We shall consider a cell similar to the one studied in Section 6.6.1, except for the electrodes. The electrodes of solid conductor A, reversible to the cation, are replaced by electrodes reversible to the anion, X_2 gas in contact with platinum metal.

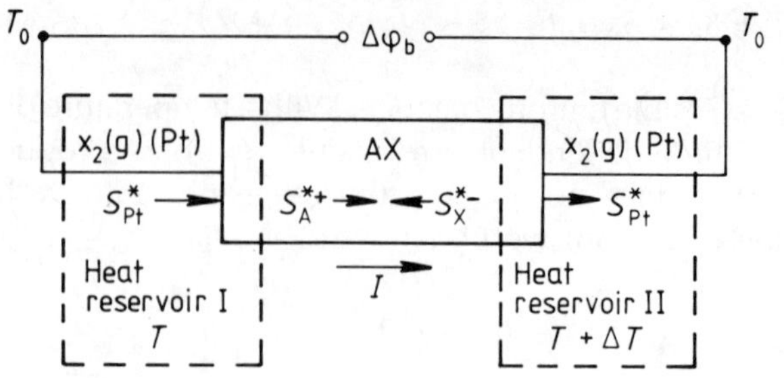

Figure 6.7. Schematic arrangement of the non-isothermal cell $X_2|AX|X_2$ (see text for explanation of the symbols).

Again we shall account the changes taking place upon the passage of one faraday of positive charge from left to right through the inner circuit of the cell. At the left-hand side one mole X^- is discharged to form one half mole X_2(g). Some of the discharged X^- is replenished by the migration of X^- up to the electrode, while the migration of A^+ away from the electrode leads to a transfer of AX to the right-hand-side electrode where X^- is formed. The total change at the left-hand-side electrode is $+\frac{1}{2}$ mol X_2 $-$ t_{A^+} mol AX. At the right-hand-side electrode, conversely, the total change is $-\frac{1}{2}$ mol X_2 $+$ t_{A^+} mol AX.

The dissipated energy per unit time and the flux equations are similar to the ones for the previous cell (see eqs. (6.33) and (6.34)). For the present system, however, we shall use the symbol $\Delta\varphi_b$ for the electric potential. Further, we shall use the symbol $\pi_{T,b}$ for the Peltier heat at the temperature T. The Peltier heat is

$$(J_q/I)_{\Delta T=0} = \pi_{T,b} \tag{6.42}$$

Similarly to the previous cell, we have the following for $I = 0$:

$$\Delta\varphi_b = -\,\pi_{T,b}\Delta \ln T = -\,(\pi_{T,b}/T)\Delta T \tag{6.43}$$

As for the previous cell we shall consider the balance of entropy across the left-hand-side electrode – electrolyte interface. The interface receives entropy from the heat reservoir, $\pi_{T,b}/T$, entropy transported through the platinum metal to the interface, S^*_{Pt} and entropy transported through the electrolyte to the interface, $t_{X^-}S^*_{X^-}$. The disappearance of t_{A^+} mole AX liberates the entropy

$t_{A^+}S_{AX}$. This balances the entropy consumed by the formation of one half mole $X_2(g)$, $\frac{1}{2}S_{X_2}$, and the entropy transported away from the interface through the electrolyte, $t_{A^+}S^*_{A^+}$. Thus the reversible entropy balance is

$$\pi_{T,b}/T + S^*_{Pt} + t_{X^-}S^*_{X^-} + t_{A^+}S_{AX} = \tfrac{1}{2}S_{X_2} + t_{A^+}S^*_{A^+} \tag{6.44}$$

Rearranging the equation and eliminating transference numbers by means of eq. (6.40) we obtain

$$\pi_{T,b}/T = -(S^*_{Pt} - S^*_{A^+}) - (S_{AX} - \tfrac{1}{2}S_{X_2})$$

Let us compare the thermoelectric power for the two cells, a and b. The outer circuits of the two cells are different, the metal A in cell a is replaced by Pt in cell b. In order to eliminate this difference, we shall use Pt only for the electrodes, not for the leads, in cell b. The Pt electrodes are welded to leads of metal A inside the heat reservoir; hence there is no change in temperature over the Pt metal. At the interface A/Pt the difference between entropy transported away from the interface and up to the interface is $S^*_{Pt} - S^*_{A}$. The S^*_{Pt} is again transported to the Pt/AX interface. Since these changes take place at constant temperature (in the same heat reservoir), the S^*_{Pt} at the two interfaces cancels. The thermoelectric power of cell b, $\Delta\varphi_b/\Delta T$, will take on the value

$$\Delta\varphi_b/\Delta T = (S^*_{A} - S^*_{A^+}) + (S_{AX} - \tfrac{1}{2}S_{X_2}) \tag{6.45}$$

Now that the difference in the outer circuit is removed, we obtain the following from eqs. (6.41) and (6.45):

$$\Delta\varphi_a/\Delta T - \Delta\varphi_b/\Delta T = S_A + \tfrac{1}{2}S_{X_2} - S_{AX} = -\Delta S \tag{6.46}$$

All entropies transported with the current cancel, and $\Delta S = S_{AX} - S_A - \frac{1}{2}S_{X_2}$ is the entropy change for the cell reaction of the isothermal cell $A|AX|X_2(Pt)$. The entropy change is equal to the temperature derivative of the emf of this cell, $d\varphi_{iso}/dT = \Delta S$, and hence we have

$$\Delta\varphi_a/\Delta T - \Delta\varphi_b/\Delta T + d\varphi_{iso}/dT = 0 \tag{6.47}$$

6.6.3 *The cell (T)A|AX,H_2O|A(T+ΔT)*

The thermocell to be discussed in this section differs from cell a by having an aqueous solution as the electrolyte instead of a pure salt. In the non-isothermal system there will then be a Soret–Dufour effect in addition to a Peltier–Seebeck effect. The cell is pictured in Fig. 6.8.

As in the treatment of the Soret effect, we shall choose water as the frame of reference. The system has three independent forces, $-\Delta\ln T$, $-\Delta\mu_{AX,T}$ and $-\Delta\varphi$ with the conjugate fluxes J_q, J_{AX} and I. In addition there is a flux $J_A = I$ with a conjugate force equal to zero. By eq. (3.22) we obtain the equation for the dissipated energy per unit time:

$$\frac{TdS}{dt} = -\Delta\ln T J_q - \Delta\mu_{AX,T} J_{AX} - \Delta\varphi I \tag{6.48}$$

The corresponding flux equations are

$$J_q = -\bar{L}_{11}\Delta\ln T - \bar{L}_{12}\Delta\mu_{AX,T} - \bar{L}_{13}\Delta\varphi \tag{6.49a}$$

$$J_{AX} = -\bar{L}_{21}\Delta\ln T - \bar{L}_{22}\Delta\mu_{AX,T} - \bar{L}_{23}\Delta\varphi \tag{6.49b}$$

$$I = -\bar{L}_{31}\Delta\ln T - \bar{L}_{32}\Delta\mu_{AX,T} - \bar{L}_{33}\Delta\varphi \tag{6.49c}$$

The electric potential of the cell, $\Delta\varphi$, is obtained from eq. (6.49c):

$$\Delta\varphi = -(\bar{L}_{31}/\bar{L}_{33})\Delta\ln T - (\bar{L}_{32}/\bar{L}_{33})\Delta\mu_{AX,T} - (1/\bar{L}_{33})I \tag{6.50}$$

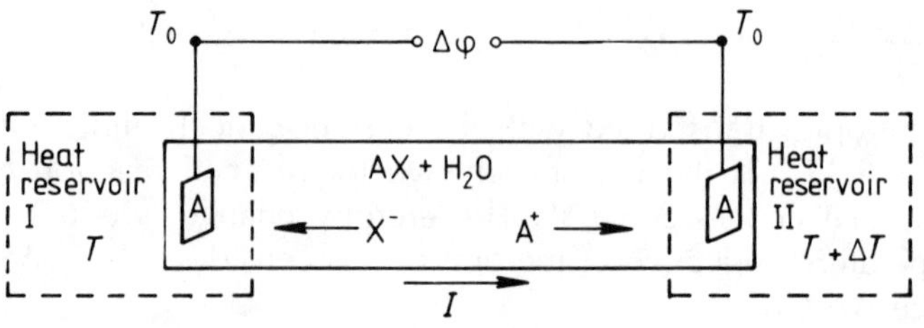

Figure 6.8. Schematic arrangement of the non-isothermal cell A|AX,H_2O|A.

In an *isothermal* system, $\Delta \ln T = 0$, the flux equations (6.49) are reduced to

$$J_{AX} = -\bar{L}_{22}\Delta\mu_{AX,T} - \bar{L}_{23}\Delta\varphi \quad (6.51a)$$

$$I = -\bar{L}_{32}\Delta\mu_{AX,T} - \bar{L}_{33}\Delta\varphi \quad (6.51b)$$

The isothermal cell is similar to the one treated in Section 4.1.2. Thus $\bar{L}_{32}/\bar{L}_{33} = \bar{L}_{23}/\bar{L}_{33} = -t_{X^-}$ (cf. eqs. (4.15 – 4.19). With no transport of neutral ion pairs we also have $\bar{L}_{22} = -\bar{L}_{23}$ (compare eq. (4.22)).

In a system with *no difference in chemical potential*, $\Delta\mu_{AX,T} = 0$, the flux equations (6.49) are reduced to

$$J_q = -\bar{L}_{11}\Delta \ln T - \bar{L}_{13}\Delta\varphi \quad (6.52a)$$

$$I = -\bar{L}_{31}\Delta \ln T - \bar{L}_{33}\Delta\varphi \quad (6.52b)$$

The ratio $\bar{L}_{31}/\bar{L}_{33} = \bar{L}_{13}/\bar{L}_{33}$ may be interpreted as the Peltier heat, π_T (cf. Section 6.6.1).

When $\Delta\mu_{AX,T} = 0$, the second term on the righthand side of eq. (6.50) vanishes. The cell is similar to cell a, treated in Section 6.6.1, and the electric potential for $I = 0$ at time zero, $t = 0$, is

$$\Delta\varphi_{t=0} = -\pi_T \Delta \ln T \approx -(\pi_T/T)\Delta T \quad (6.53)$$

This equation is the same as eq. (6.37), and π_T/T may be expressed as in eq. (6.39). This gives the thermoelectric power

$$\Delta\varphi_{t=0}/\Delta T = S_A + S_A^* - (t_{X^-}S_{AX} + t_{A^+}S_{A^+}^* - t_{X^-}S_{X^-}^*) \quad (6.54)$$

In the course of time the thermodiffusion and a current through the cell will create a difference in chemical potential across the electrolyte. By eliminating $\Delta\varphi$ from eqs. (6.49) we obtain

$$J_q = -\bar{l}_{11}\Delta \ln T - \bar{l}_{12}\Delta\mu_{AX,T} + \pi_T I \quad (6.55a)$$

$$J_{AX} = -\bar{l}_{21}\Delta \ln T - \bar{l}_{22}\Delta\mu_{AX,T} - t_{X^-}I \quad (6.55b)$$

A *heat of transfer*, q_{AX}^*, may be defined similarly to q_2^* (cf. eq. (6.10)):

$$q^*_{AX} = (J_q/J_{AX})_{\Delta T=0, I=0} = \bar{l}_{12}/\bar{l}_{22} \tag{6.56}$$

In a stationary state, where $J_{AX} = 0$, we have

$$\Delta\mu_{AX,T} = -(\bar{l}_{21}/\bar{l}_{22})\Delta \ln T - (t_{X^-}/\bar{l}_{22})I$$
$$= -q^*_{AX}\Delta \ln T - (t_{X^-}/\bar{l}_{22})I \tag{6.57}$$

For a given value of $\Delta \ln T$, a stationary state, $J_{AX}=0$, can be found for any value of $\Delta\mu_{AX,T}$ by adjusting I. At the time $t = 0$ when $\Delta\mu_{AX,T} = 0$, a substantial current must pass through the cell to counterbalance the temperature difference and obtain a stationary state. We may allow pure thermodiffusion to take place, e.g. with an open circuit. The fluxes are then represented by eqs. (6.55) with $I = 0$. The difference in chemical potential created across the electrolyte will partly counterbalance the temperature difference and, after some time, a smaller current is sufficient to obtain a stationary state, $J_{AX} = 0$. We may allow the pure thermodiffusion to go on until the stationary state is obtained where $J_{AX} = 0$ without any current, $I = 0$, at the time $t = \infty$.

The electric potential at any time can be expressed by eq. (6.50) where $\bar{L}_{31}/\bar{L}_{33}$ is replaced by π_T, $\bar{L}_{32}/\bar{L}_{33}$ by $-t_{X^-}$ and $\Delta\mu_{AX,T}$ by the expression given in eq. (6.57). With the approximation $\Delta \ln T \approx \Delta T/T$ we obtain

$$\Delta\varphi = -(\pi_T/T)\Delta T - t_{X^-}(q^*_{AX}/T)\Delta T - t_{X^-}(t_{X^-}/\bar{l}_{22})I - (1/\bar{L}_{33})I \tag{6.58}$$

As for cell a, treated in Section 6.6.1, we may consider the entropy balance across the metal – electrolyte interface for the left-hand-side electrode when passing one faraday of positive charge from left to right through the inner circuit of the cell. We obtain expressions which are the same as eqs. (6.38) and (6.39). The meaning of some of the symbols, however, is slightly different; S_{AX} is here the partial molar entropy of AX in aqueous solution and $S^*_{A^+}$ and $S^*_{X^-}$ represent entropies transported through an aqueous solution.

$$\pi_{T,a}/T + S^*_A + S_A + t_{X^-}S^*_{X^-} = t_{X^-}S_{AX} + t_{A^+}S^*_{A^+} \tag{6.38}$$

$$\pi_{T,a}/T = -S_A - S_A^* + (t_{X^-}S_{AX} + t_{A^+}S_{A^+}^* - t_{X^-}S_{X^-}^*) \tag{6.39}$$

From eqs. (6.58) and (6.39) we obtain the thermoelectric power

$$\Delta\varphi/\Delta T = (S_A + S_A^*) - (t_{X^-}S_{AX} + t_{A^+}S_{A^+}^* - t_{X^-}S_{X^-}^* + t_{X^-}q_{AX}^*/T) - (t_{X^-}^2/\bar{l}_{22} + 1/\bar{L}_{33})I/\Delta T \tag{6.59}$$

This equation is valid for stationary state, where $J_{AX} = 0$. Then there is no net movement of X^- and the entropy changes in the electrolyte at the interface are the same as those caused by the transport of one mol A^+ away from the interface:

$$t_{X^-}S_{AX} + t_{A^+}S_{A^+}^* - t_{X^-}S_{X^-}^* + t_{X^-}q_{AX}^*/T = S_{A^+}^* \tag{6.60}$$

Rearranging the equation and remembering that $t_{A^+} + t_{X^-} = 1$, we obtain

$$S_{AX} + q_{AX}^*/T = S_{A^+}^* + S_{X^-}^* \tag{6.61}$$

All the terms in the equation are independent of forces, and the equation is valid whether we have the stationary state or not.

From eqs. (6.59) and (6.60) we obtain a simple expression for the thermoelectric power in the stationary state

$$\Delta\varphi/\Delta T = S_A + S_A^* - S_{A^+}^* - (t_{X^-}^2/\bar{l}_{22} + 1/\bar{L}_{33})I/\Delta T \tag{6.62}$$

At the time $t = \infty$, when $J_{AX} = 0$ for $I = 0$, we have

$$\Delta\varphi_{t=\infty}/\Delta T = S_A + S_A^* - S_{A^+}^* \tag{6.63}$$

At the time $t = 0$, the expression for the thermoelectric power is obtained from eqs. (6.54) and (6.60):

$$\Delta\varphi_{t=0}/\Delta T = S_A + S_A^* - S_{A^+}^* + t_{X^-}q_{AX}^*/T \tag{6.64}$$

Since it takes a very long time to reach the stationary state, eq. (6.64) for the thermoelectric power will usually give a better approximation to reality than eq. (6.63). Note that *absolute* entropies are needed for calculations of thermoelectric power. This is unlike calculations in equilibrium thermodynamics, where only entropy differences are needed.

We shall return to eq. (6.61). It is customary to split S_{AX} into separate terms for the formation of single ions, $S_{AX} = S_{A^+} + S_{X^-}$. Similarly the entropy transported by diffusion is split into single ion diffusion entropies, $q^*_{AX}/T = \hat{S}_{A^+} + \hat{S}_{X^-}$, the *Eastman entropies*.[6] Hence we have,

$$S^*_{A^+} = S_{A^+} + \hat{S}_{A^+} \tag{6.65a}$$

$$S^*_{X^-} = S_{X^-} + \hat{S}_{X^-} \tag{6.65b}$$

None of these single ion functions, S_{A^+}, S_{X^-}, $\hat{S}_{A^+}$ or $\hat{S}_{X^-}$, can be obtained as an experimental value.

In this chapter we have dealt with transports in systems with temperature gradients combined with other forces. Methods of calculation for simple systems have been presented. The extension to more complex systems does not represent anything new in principle. A temperature gradient influences the transport of mass and charge. It may create gradients in pressure, concentration and electric potential. In many cases these effects may be substantial.

6.7 *References*

1. Harman, T. C. and Honig, J. M., *Thermoelectric and Thermomagnetic Effects and Applications*, McGraw-Hill, New York, 1967.
2. de Groot, S. R. and Mazur, P., *Non-equilibrium Thermodynamics*, North-Holland, Amsterdam, 1962.
3. Agar, J. N., Thermogalvanic cells, in *Electrochemistry and Electrochemical Engineering* (ed. Delaney, P.), Interscience, New York, 1963.
4. Sundheim, B. R., Transport properties of liquid electrolytes, in *Fused Salts* (ed. Sundheim, B. R.), McGraw-Hill, New York, 1964.
5. Katchalsky, A. and Curran, P. F., *Nonequilibrium Thermodynamics in Biophysics*, Harvard University Press, 1965.
6. Eastman, E. D., *J. Am. Chem. Soc.*, **50**, 283 (1928).

6.8. *Exercises*

6.1. *The Soret effect.* A cylindrical container of length 5 mm is filled with an aqueous solution of KCl, c_{KCl} = 0.01 kmol m^{-3}.

The temperature on the left-hand side, ℓ, is 20°C, and on the right-hand side, r, 30°C. In the stationary state a difference in concentration is established between the two sides, $c_{KCl,r} - c_{KCl,\ell} = \Delta c_{KCl} = -2.5\times10^{-4}$ kmol m^{-3}. The average Fick's diffusion coefficient for KCl in water is $D = 1.9\times10^{-9}$ m^2 s^{-1}. (a) Calculate the *Soret coefficient*, s_T, and the *heat of transfer*, q^* (use the mean temperature), for KCl in the solution. (b) Calculate the flux of KCl at the starting time, when $\Delta c_{KCl} = 0$.

6.2. *The Thomson coefficient*. The entropy transport in platinum metal was investigated using an experimental arrangement like the one shown in Fig. 6.3. Metal B was Pt while metal A was Pb. The entropy transport in Pb, S^*_{Pb}, is known with great accuracy. The result obtained for S^*_{Pt} was

$$S^*_{Pt} = (-\,0.8620 - 0.001146\,T + 229\,T^{-1} - 8772\,T^{-2})$$

$$\text{J K}^{-1}\ \text{faraday}^{-1}$$

Find the Thomson coefficient for Pt.

6.3. *A thermocell*. Consider the thermocell, where $c_{KCl} = 0.01$ kmol m^{-3}, (T_1)Ag(s)|AgCl(s)|KCl(aq)|AgCl(s)|Ag(s)(T_2). (a) Study the reversible entropy balance across the metal – electrolyte interface at the left-hand-side electrode and find an expression for the thermoelectric power at the starting time when the electrolyte has a uniform composition. (b) Calculate the entropy transferred through the electrolyte, $t_{K^+}S^*_{K^+} - t_{Cl^-}S^*_{Cl^-}$, from the following data at 25°C. The thermoelectric power, $\Delta\varphi/\Delta T = 57.9$ J K^{-1} faraday^{-1} for $t = 0$. The partial molar entropy for KCl, $S_{KCl} = 236.7$ J K^{-1} mol^{-1} for $c_{KCl} = 0.01$ kmol m^{-3}. Further, $S_{Ag(s)} = 42.7$ J K^{-1} mol^{-1}, $S_{AgCl(s)} = 96.2$ J K^{-} mol^{-1}. The ionic transference numbers, $t_{K^+} = t_{Cl^-} = 0.5$. The entropy transferred in metallic leads can be neglected. The temperature difference is small, and the average temperature is 25°C. (c) Use the result from Exercise 6.1a to calculate $\Delta\varphi/\Delta T$ for $t = \infty$, $S^*_{K^+}$ and $S^*_{Cl^-}$.

CHAPTER 7

Systems in the Gravitational Field. The Ultracentrifuge

So far we have not considered the change in internal energy of a system with height. The force of gravity on a particle is proportional to its mass and this should in principle lead to some separation of mixtures with altitude. For most homogeneous systems the limitation in height renders the influence of the gravitational field of the earth on the fluxes entirely negligible.

For gaseous mixtures we are not likely to find any observable effect. The average molar mass of the atmosphere of the earth is constant up to about 90 km above sea level, as any demixing effect of gravity is completely effaced. One might expect that gravity would have somewhat more influence on liquid mixtures. For heterogeneous mixtures the sedimentation of the denser particles due to the gravitational force is the over-riding influence, and the phenomenon can be treated without resorting to irreversible thermodynamics.

By means of a centrifuge one can obtain accelerations much higher than the acceleration of free fall. The *relative centrifugal force* (RCF), the ratio of the acceleration of the centrifuge to the acceleration of free fall, can reach several hundred thousand. Then the force of gravity will be of prime importance.

The gas centrifuge is used commercially to separate isotopes of uranium by centrifugation of UF_6 gas. The difference in density between the fluorides of the isotopes is very small and the degree of separation in one centrifuge unit is far less than desired. By connecting centrifugal units in series (cascade separation) a satisfactory degree of separation can be obtained.

A liquid centrifuge is used for separating dissolved substances from homogeneous solution and for separating phases from a suspension of solids in a liquid phase. The efficiency of the separation depends on the density, size and shape of the dissolved molecules and the suspended particles, and on the RCF of the centrifuge. An *ultracentrifuge* has a high RCF and can give separations down to molecular dimensions. A *preparative* ultracentrifuge is used in biology to concentrate or separate cells, subcellular organelles or large molecules from water; it is also used in chemistry to concentrate or separate macromolecules (high polymers) from the solvent. The *analytical* ultracentrifuge is a more advanced instrument with an accurate temperature control and optics that permit the study of the sedimentation process during the centrifugation. By means of an analytical ultracentrifuge one may obtain information on molar masses, buoyant densities, sedimentation coefficients and diffusion coefficients.

In this chapter we shall discuss the basic principles governing systems in a gravitational or centrifugal field.

7.1 *The Chemical Potential in a Gravitational or Centrifugal Field*

In the absence of any field of acceleration, the chemical potential of a species is a function of composition, temperature and pressure. Under the influence of the gravitational field, the chemical potential is in addition a function of the level in the gravitational field:

$$\mu_i = \mu_i\,(c,T,p) + \textit{gravitational potential} \qquad (7.1)$$

The energy contribution from the gravitational field to a mass m_i of component i at the level (or altitude) h is $m_i gh$, where g is the acceleration of free fall. The *gravitational potential* is the energy contribution per mole i:

$$\textit{gravitational potential} = (m_i/n_i)gh = M_i gh \qquad (7.2)$$

where M_i is the molar mass of component i.

The gradient in chemical potential in the direction of *increasing* gravitational potential is

$$\mathrm{d}\mu_{i,T}/dh = \mathrm{d}\mu_{i,T}(c)/dh + \mathrm{d}\mu_{i,T}(p)/dh + M_i g dh/dh \qquad (7.3)$$

The gradient in gravitational potential is $M_i g$, while the gradient in chemical potential caused by a pressure gradient is given by eq. (5.16), $d\mu_{i,T}(p)/dh = V_i dp/dh$. The change of pressure with altitude is determined by the density of the medium, ρ, $dp/dh = -\rho g$. Hence we have

$$d\mu_{i,T}(p)/dh = -V_i \rho g \tag{7.4}$$

The forces, X_i, can be expressed as follows:

$$X_i = -d\mu_{i,T}/dh = -d\mu_{i,T}(c)/dh - (M_i - V_i\rho)g \tag{7.5}$$

When there is a temperature gradient in the system, there will in addition be the force $X_q = -d\ln T/dh$, and in an electric field the additional force $X_j = -d\varphi/dh$ (cf. eq. (2.27)).

An expression similar to eq. (7.5) for any accelerating field is obtained by replacing g by a, a general symbol for acceleration. Then h is the distance in the accelerating field. In a centrifuge the acceleration is $a = \omega^2 x$, where ω is the angular velocity of the centrifuge (radians per second) and x is the distance from the centre of rotation. The distance from the centre is the distance in the direction of *decreasing* centrifugal potential, and the forces, X_i, in a centrifugal field are

$$X_i = -d\mu_{i,T}/dx = -d\mu_{i,T}(c)/dx + (M_i - V_i\rho)\omega^2 x \tag{7.6}$$

The very small Coriolis force is neglected. It is common to introduce the *specific volume*, $v_i = V_i/M_i$, in the equation giving

$$X_i = -d\mu_{i,T}/dx = -d\mu_{i,T}(c)/dx + M_i(1 - v_i\rho)\omega^2 x \tag{7.7}$$

When an isothermal system has attained equilibrium, $X_i = 0$ for all components and for each component we have,

$$d\mu_i(c)/dx = M_i(1 - v_i\rho)\omega^2 x \tag{7.8}$$

This equation may be integrated over a distance x_I to x_{II} with a corresponding concentration range $c_{i,I}$ to $c_{i,II}$. Assuming v_i and ρ constant we obtain

$$\mu_i(c)_{II} - \mu_i(c)_I = \tfrac{1}{2} M_i(1 - v_i\rho)\omega^2(x_{II}^2 - x_I^2)$$

For an ideal solution we have $\mu_i(c)_{II} - \mu_i(c)_I = RT\ln(c_{i,II}/c_{i,I})$, and we obtain the following expression for the molar mass:

$$M_i = \frac{2\,RT\ln(c_{i,II}/c_{i,I})}{\omega^2(1 - v_i\rho)(x_{II}^2 - x_I^2)} \tag{7.9}$$

When the concentration of a component is known as a function of distance, the molar mass of the component can be found. When the molar mass is known, the chemical potential as a function of composition can be found for a non-ideal solution. In an analytical ultracentrifuge the concentration of a component as a function of distance can be determined by observing the refractive index of the solution as a function of distance.

A major drawback of this method is the long time required to reach equilibrium. It is therefore of interest to investigate the non-equilibrium situation with fluxes of components. For this we shall apply the methods of irreversible thermodynamics.

7.2 *Flux Equations for Systems in a Centrifugal Field*

The analytical ultracentrifuge is commonly used in a situation of non-equilibrium, where particles move under the influence of centrifugal acceleration and concentration gradients. The movements are recorded and interpretations of these movements are derived by means of flux equations.

In order to simplify derivations we have chosen a system consisting of a solvent, 1, and a solute, 2. We shall treat an isothermal system with no electric current. Three different frames of reference will be used. To distinguish between them superscripts will be used on the flux symbols, and, when needed, also on other symbols. The fluxes, J^S, J^V and J^C refer to solvent-fixed, volume-fixed and cell-fixed frames of reference respectively.

7.2.1 *Solvent-fixed frame of reference*

For the *solvent-fixed* frame of reference, fluxes are measured relative to the solvent, and $J_1^S = 0$. Our two-component system has one independent force, $-\nabla\mu_2$, and one flux, J_2^S, the flux of the solute. The flux equation is

$$J_2^S = -l\,\nabla\mu_2 \tag{7.10}$$

where l is a diffusion coefficient. The flux is considered in the x-direction only (Coriolis force is neglected). Then the force is given by eq. (7.7) and we obtain

$$J_2^S = lM_2(1 - v_2\rho)\omega^2 x - l\mathrm{d}\mu_{2,T}(c)/\mathrm{d}x \tag{7.11}$$

Commonly a *sedimentation coefficient*, $\mathcal{S}^S$, is introduced. It is defined by the equation

$$\mathcal{S}^S = lM_2(1 - v_2\rho)/c_2 \tag{7.12}$$

The velocity of component 2 relative to the solvent, 1, is $\boldsymbol{v} = J_2^S/c_2$, and we can see from eqs. (7.11) and (7.12) that $\mathcal{S}^S$ is the relative velocity of component 2 per unit centrifugal force field ($\omega^2 x = 1$) when $\mathrm{d}\mu_{2,T}(c)/\mathrm{d}x = 0$.

For an ideal mixture we have (cf. eq. (4.27b))

$$l\mathrm{d}\mu_{2,T}(c)/\mathrm{d}x = \frac{lRT}{c_2}\mathrm{d}c_2/\mathrm{d}x = D^S\mathrm{d}c_2/\mathrm{d}x \tag{7.13}$$

where D^S is the (solvent-fixed) *Fick's diffusion coefficient*. When there is no centrifugal force field in the system, eq. (7.11) reduces to $J_2^S = -D^S\mathrm{d}c_2/\mathrm{d}x$, Fick's first law of diffusion (cf. eq. (4.28)).

Introducing eqs. (7.12) and (7.13) into eq. (7.11) we obtain

$$J_2^S = \mathcal{S}^S c_2\omega^2 x - D^S\mathrm{d}c_2/\mathrm{d}x \tag{7.14}$$

A very important use of ultracentrifugation is the determination of molar mass by the combined measurement of sedimentation coefficients and Fick's diffusion coefficients. In an experimental situation, however, the solvent-fixed frame of reference is not very convenient. A *cell-fixed* frame of reference is used. If there is no volume change during the experiment, it is identical to a *volume-fixed* frame of reference.

7.2.2 *Volume-fixed frame of reference*

We shall derive relations between the fluxes J_2^S and J_2^V using concepts discussed in Section 5.4.2. A volume-fixed frame of reference means that the volume flux, J_V, is equal to zero. Thus we have (cf. eq. (5.39)):

$$J_V^V = J_1^V V_1 + J_2^V V_2 = 0 \tag{7.15}$$

The difference in velocity for components 2 and 1, $v_2 - v_1$, is the diffusion flux, J_D^V (cf. eq. (5.40)):

$$J_D^V = J_2^V/c_2 - J_1^V/c_1 \tag{7.16}$$

The flux of the solute relative to the solvent, J_2^S, is equal to the diffusion flux multiplied by the concentration of the solute:

$$J_2^S = c_2 J_D^V = J_2^V - J_1^V c_2/c_1 \tag{7.17}$$

We eliminate J_1 from eq. (7.17) by combining it with eq. (7.15):

$$J_2^S = J_2^V (1 + c_2 V_2/c_1 V_1)) \tag{7.18}$$

Since $c_1 V_1 + c_2 V_2 = 1$, we obtain

$$J_2^S = J_2^V \frac{1}{c_1 V_1} \tag{7.19}$$

Introducing eq. (7.19) into eq. (7.14) we obtain

$$J_2^V = (\mathcal{S}^S c_1 V_1) c_2 \omega^2 x - (D^S c_1 V_1) dc_2/dx \tag{7.20}$$

or

$$J_2^V = \mathcal{S}^V c_2 \omega^2 x - D^V dc_2/dx \tag{7.21}$$

where $\mathcal{S}^V$ and D^V refer to a volume-fixed frame of reference.

For dilute solutions $c_1 V_1 \approx 1$, which gives

$$J_2^V \approx J_2^S\,; \quad \mathcal{S}^V \approx \mathcal{S}^S\,; \quad D^V \approx D^S \tag{7.22}$$

The advantage of using a solvent-fixed frame of reference, rather than a volume-fixed one, is that flux equations with independent forces are directly obtained when one force is eliminated by the use of the Gibbs–Duhem equation relating the chemical potentials. This becomes more important when there are several components.

For comparison of experimental determinations, the volume-fixed frame of reference is more convenient. For practical purposes this is the same as the cell-fixed one.

7.2.3 *Cell-fixed frame of reference*

For the cell-fixed frame of reference, fluxes are measured relative to the walls of the container. One may, for example, refer all movements to the closed end of a cylindrical cell.

If the partial molar volumes of the components vary with composition, the transport will be accompanied by a volume change. As a consequence there will be a net volume flow, $J_V \neq 0$, relative to the walls of the container. In liquid mixtures, however, relative changes in partial molar volumes are usually very small. Hence the cell-fixed and the volume-fixed frames of reference are practically identical.

A detailed discussion of different frames of reference giving equations for the conversion of fluxes from one frame of reference to another is given by Sundheim[1] (see also discussions by Kirkwood and coworkers[2]).

7.3 *Sedimentation Velocity from Stoke's Law*

Stoke's law states that the force, X, needed to draw a spherical particle of radius, r, with a velocity, v, through a medium of viscosity, η, is given by

$$X = 6\pi\eta r v \tag{7.23}$$

For non-spherical particles the expression should be multiplied by a shape factor.

Stoke's law, as given here, implies a solvent-fixed frame of reference. In a two-component system the solvent, component 1, is the medium, and the force is given by the gradient in chemical potential for component 2. Referring to a single particle of the solute, the force is $X = -\nabla\mu_2/N_A$, where N_A is Avogadro's constant. The velocity of component 2 relative to the solvent is $v = J_2^S/c_2$. Hence the diffusion coefficient, l, of eq. (7.10) is

$$l = \frac{c_2}{6\pi\eta r N_A} \tag{7.24}$$

Introducing this expression for l in eqs. (7.12) and (7.13) expressions are obtained for $\mathcal{S}^S$ and D^S:

$$\mathcal{S}^S = \frac{(1 - v_2\rho)M_2}{6\pi\eta r N_A} \tag{7.25}$$

and

$$D^S = \frac{RT}{6\pi\eta r N_A} \tag{7.26}$$

The ratio between the two is the *Svedberg equation*:

$$\mathcal{S}^S/D^S = \frac{(1 - v_2\rho)M_2}{RT} \tag{7.27}$$

Experimental values for $\mathcal{S}^S$ are found by the study of the rate of sedimentation in an ultracentrifuge. According to eq. (7.14) D^S must be known in order to find $\mathcal{S}^S$. If ω^2x is made sufficiently large, the second term on the right-hand side of the equation may be negligible compared to the first term. The value of D^S can be determined by a separate diffusion experiment. The molar mass of component 2, M_2, can be found from eq. (7.27) when the density of the solution, ρ, and the specific volume of the solute, v_2, are known. The radius of the molecules of component 2, r, can be found from eq. (7.25) or (7.26) when the viscosity of the solution, η, is known.

7.4 Sedimentation in a Ternary Mixture

We have a ternary mixture of components 1, 2 and 3. Component 1 may be considered as the solvent. There are two independent forces in the system and, with a solvent frame of reference, the forces can be expressed by eq. (7.7):

$$X_2 = -\,d\mu_{2,T}/dx = -\,d\mu_{2,T}(c)/dx + M_2(1 - v_2\rho)\omega_2 x \tag{7.28a}$$

$$X_3 = -\,d\mu_{3,T}/dx = -\,d\mu_{3,T}(c)/dx + M_3\,(1 - v_3\rho)\omega^2 x \tag{7.28b}$$

For an isothermal system the subscript T may be omitted, and we can write $d\mu_i$ instead of $d\mu_{i,T}$. The flux equations for the two solutes are

$$J_2^S = -l_{22}(d\mu_2/dx) - l_{23}(d\mu_3/dx) \tag{7.29a}$$

$$J_3^S = -l_{32}(d\mu_2/dx) - l_{33}(d\mu_3/dx) \tag{7.29b}$$

Introducing the expressions for the forces (eqs. (7.28)) into eqs. (7.29) will give rather complex functions for the fluxes. When expressing the fluxes by sedimentation coefficients, $\mathcal{S}^S$, and Fick's diffusion coefficients, D^S, the functions will also be complex. The sedimentation coefficients will contain terms for both solutes and there will be four diffusion coefficients, D_{22} and D_{33} representing the diffusion relative to the solvent and D_{23} and D_{32} representing interaction between the two solutes. The problem is discussed in more detail by Katchalsky and Curran.[3]

In a homogeneous solution of the components 2 and 3 in the solvent, component 1, $d\mu_2(c)/dx = 0$ and $d\mu_3(c)/dx = 0$, and the forces given by eqs. (7.28) are reduced to respectively $X_2 = M_2(1 - v_2\rho)$ and $X_3 = M_3\ (1 - v_3\rho)$. The direction of a force is then determined by the value of $v_i\rho$. If $v_i\rho < 1$, the component will migrate towards the bottom of the centrifuge cell (away from the centre of rotation); conversely, if $v_i\rho > 1$, the component will migrate towards the top of the cell (for the solvent we have $v_1\rho = 1$). This means that two compounds with different specific volume can be separated effectively from a mixture by centrifugation, if a solvent with a value for specific volume between the two is used.

A solution of varying density may be used as the medium in the centrifuge cell, where a third compound is sedimented. For dilute solutions a constant value for ρ may be assumed. For more concentrated solutions, however, there may be a substantial variation of ρ with concentration. If, for example, sucrose or caesium chloride in aqueous solution is used as a medium in the centrifuge cell, the density and the concentration of the solution will increase with the distance from the centre of rotation. A third component will not migrate all the way to the bottom of the cell (or to the top of the cell), but to the level where $v_3\rho = 1$. At this point there will be a peak in the concentration of component 3, the width of the peak being determined by diffusion. In such a density gradient cell one may separate several compounds, each one having a peak in concentration at the point where $v_i \rho = 1$.

The ultracentrifuge is a very useful instrument for many different separation purposes. A thorough treatment has been given by Hsu.[4]

7.5 *References*

1. Sundheim, B. R., Transport properties of liquid electrolytes, in *Fused Salts* (ed. Sundheim, B. R.), McGraw-Hill, New York, 1964.
2. Kirkwood, J. G., Baldwin, R. L., Dunlop. P. J., Gosting, L. J. and Kegeles, G., *J. Chem. Phys.*, **33**, 1505 (1960).
3. Katchalsky, A. and Curran, P. F., *Nonequilibrium Thermodynamics in Biophysics*, Harvard University Press, 1965.
4. Hsu, H. W., Separation by centrifugal phenomena, in *Techniques of Chemistry* (ed. Weissberger, A.), vol XVT, Wiley, New York, 1981.

7.6 *Exercises*

7.1. *Dissipation function in a centrifugal field.* A liquid mixture of two components, 1 and 2, is exposed to a centrifugal field. Temperature is constant. (a) Use a volume-fixed frame of reference and give the equation for the dissipation function. What is the contribution to the dissipation function from the centrifugal field? (b) The forces in (a) are mutually dependent, and so are the fluxes. Apply the Gibbs–Duhem equation and express the dissipation function by one flux and one force. Give the flux equation.

PART II
Applications

CHAPTER 8

The Electromotive Force of Cells with Liquid Junctions and with Membranes

When an electrochemical cell combines two electrodes operating in different solutions, the two half-cells are connected by means of a *liquid junction* or a *membrane* to prevent mixing of the solutions.

The transport of ions in a liquid junction contributes to the emf of the cell. In the study of cells with liquid junctions approximate calculations are frequently used to evaluate this contribution. A *salt bridge* between the electrolytes is used to 'eliminate', or rather to reduce substantially, the *liquid junction potential*.

Transport of ions in a membrane will also contribute to the emf of a cell. We shall see that the selectivity of a membrane can be decisive for the kind of concentration function that the emf of a cell will be. Membranes that strongly favour the transport of one particular ion form the basis for *ion-selective electrodes*. An ion-selective electrode is used as an indicator electrode in a test solution connected via a salt bridge to a reference electrode. The emf of the cell is then a direct function of the concentration of the particular ion in the test solution.

A glass electrode with an H^+-selective glass membrane was already known at the beginning of the century. It is still one of the most commonly used ion-selective electrodes. Ion-selective electrodes for other ions were developed later. During the last 20–30 years there has been an increasing interest in the field of ion-selective electrodes. As a result of intensive research, electrodes specific to a number of different cations and anions have been made available. Further development with new electrodes and improvements in present ion-selective electrodes can be expected. We shall discuss the significant parameters for the selectivity of an electrode.

Ion-selective electrodes are used widely. For the analytical chemist they offer a rapid method to determine the concentration of a specific ion in a solution. They can be used for monitoring industrial processes and in pollution control. In biochemistry and medicine ion-selective electrodes have many and varied applications.

Cells containing combinations of ion exchange membranes can be used for storage of energy. In such a cell the main contribution to the emf comes from the chemical reaction in the electrolyte between an anion exchange membrane and a cation exchange membrane. The redox reaction at the electrodes gives only a minor contribution to the emf.

8.1 *The Liquid Junction and the Salt Bridge*

When two half-cells containing different electrolytes are combined, an interdiffusion between the two electrolytes will take place. In order to prevent or delay the mixing of the electrolytes, the diffusion path is made narrow and long in the *liquid junction* between the electrolytes. We shall consider two kinds of liquid junction, that between two electrolytes with a common anion, both of the same concentration, and that between a dilute solution of HCl and a KCl solution of high concentration (4 kmol m^{-3}). The latter one is commonly used in a *salt bridge*.

For electrochemical cells with liquid junctions, part of the emf originates from processes in the junction, the *liquid junction potential*. We wish to evaluate the contribution from these processes. The basic equations for liquid junctions were derived in Chapter 4. For a three-component system the flux equations are similar to eqs. (4.54). In addition to the Onsager reciprocal relations, some other important relations between coefficients were derived, e.g. $L_{11} + L_{12} = L_{13}$ (eq. (4.58)). When the cross coefficient, L_{12}, has a small numerical value compared to the other two coefficients, $L_{11} \approx L_{13}$. This is the basis for the Nernst–Planck flux equations. We shall discuss the validity of these relations.

In order to calculate the contribution to the emf of a cell from the liquid junction, knowledge of concentration profiles, mobility of ions and activities of components is required. These are usually not known in detail, and assumptions have to be made in order to carry out the calculations. Approximate methods involving simplifying assumptions will be used and the results will be compared with those of more accurate calculations and with experimental values.

A salt bridge between electrolytes is used to reduce the contribution to the emf from a liquid junction. We shall carry out some approximate calculations on the effect of the salt bridge.

8.1.1. *Cells with liquid junctions*

The emf of the following cell a will be studied:

$$\underset{}{Ag(s)|AgCl(s)|}\underset{I}{HCl(aq,c_o)}||\underset{II}{KCl(aq,c_o)}|AgCl(s)|Ag(s) \tag{8.1}$$

The electrolyte on the left-hand side of the liquid junction, electrolyte I, contains an aqueous solution of HCl, while the electrolyte on the right-hand side, electrolyte II, contains an aqueous solution of KCl. The two solutes with the common anion, Cl^-, are of the same concentration, c_o, and the electrodes are identical and reversible to the common anion. Hence the emf of the cell is created in the liquid junction.

The emf of the cell is given by an equation similar to eq. (4.87):

$$\Delta\varphi = -\int_{(I)}^{(II)} (t_{HCl}\mathrm{d}\mu_{HCl} + t_{KCl}\mathrm{d}\mu_{KCl}) \tag{8.2}$$

With electrodes reversible to Cl^- we have $t_{HCl} = t_{H^+}$ and $t_{KCl} = t_{K^+}$ (cf. eqs. (4.55) and (4.56)). The transference numbers, t_{H^+} and t_{K^+}, are functions of the mole fractions of H^+ and K^+, respectively, i.e. they are functions of the concentration profiles in the liquid junction. They are also functions of the mobilities of the ions. Chemical potentials can be expressed by concentrations and activity coefficients.

Without detailed knowledge about concentration profiles, mobilities and activities, one can still calculate an approximate value for the emf after making some simplifying assumptions. We shall first use an *approximate method* for calculating the emf. We shall proceed with more accurate *computer calculations*. Then we shall compare the calculated values to *experimentally determined values*. Finally we shall discuss *other methods* for calculating emf.

Approximate method for calculating the emf of cell a. We shall base the calculations on the following assumptions: *activity coefficients are constant* across the junction, *mobilities are constant* across the junction and *concentration of* Cl^- is constant ($c_{Cl^-} = c_o$) across the junction. No assumption is made about any particular set of concentration profiles.

The ionic transference numbers can be expressed by mobilities, u_i, and concentrations, c_i, or mole fractions, $x_i = c_i/c_o$:

$$t_{H^+} = x_{H^+}u_{H^+}/\Sigma x_i u_i \tag{8.3a}$$

$$t_{K^+} = x_{K^+}u_{K^+}/\Sigma x_i u_i \tag{8.3b}$$

$$t_{Cl^-} = u_{Cl^-}/\Sigma x_i u_i \tag{8.3c}$$

The chemical potential of HCl is

$$\mu_{HCl} = \mu^o_{HCl} + RT \ln (c_{H^+}c_{Cl^-}y_{HCl}) \tag{8.4}$$

The temperature is constant and c_{Cl^-} and y_{HCl} are assumed to be constant. Thus c_{H^+} is the only variable on the righthand side of the equation. Using the mole fraction x_{H^+} the following derivative is obtained:

$$d\mu_{HCl} = RT dx_{H^+}/x_{H^+} \tag{8.5}$$

Similarly we obtain

$$d\mu_{KCl} = RT dx_{K^+}/x_{K^+} \tag{8.6}$$

Substituting eqs. (8.3), (8.5) and (8.6) into eq. (8.2) we obtain

$$\Delta\varphi_a = -RT\int_{(I)}^{(II)} \{(u_{H^+}/\Sigma x_i u_i)\, dx_{H^+} + (u_{K^+}/\Sigma x_i u_i)\, dx_{K^+}\} \tag{8.7}$$

We have $x_{H^+} + x_{K^+} = 1$ and $x_{Cl^-} = 1$.
Thus $dx_{K^+} = -dx_{H^+}$ and $\Sigma x_i u_i = x_{H^+}u_{H^+} + x_{K^+}u_{K^+} + x_{Cl^-}u_{Cl^-} = (u_{Cl^-} + u_{K^+}) + (u_{H^+} - u_{K^+})x_{H^+}$.
Thus we obtain

$$\Delta\varphi_a = -RT\int_1^0 \frac{(u_{H^+} - u_{K^+})\, dx_{H^+}}{u_{Cl^-} + u_{K^+} + (u_{H^+} - u_{K^+})x_{H^+}}$$

or

$$\Delta\varphi_a = RT \ln \{(u_{H^+} + u_{Cl^-})/(u_{K^+} + u_{Cl^-})\} \tag{8.8}$$

According to this equation the emf of the cell is independent of the value of c_o.

Using the numerical values for the mobilities at infinite dilution at 25°C,[1] $u_{H^+} = 36.3 \times 10^{-8}$ ms^{-1}/Vm^{-1}, $u_{K^+} = 7.61 \times 10^{-8}$ ms^{-1}/Vm^{-1} and $u_{Cl^-} = 7.91 \times 10^{-8}$ ms^{-1}/Vm^{-1}, we obtain

$$E = \Delta\varphi_a/F = 26.9 \text{ mV} \tag{8.9}$$

A correction for variation in the activity coefficient can be introduced into eq. (8.4) by the use of *Harned's rule*:[2] the activity coefficient of HCl in the mixture is given by the empirical equation

$$\ln y_{HCl} = \ln y^{o}_{HCl} + Kc_{K^+} \tag{8.10}$$

where $\ln y^{o}_{HCl}$ represents the pure HCl solution and K is a constant for constant ionic strength in the solution. A similar correction can be introduced in the expression for μ_{KCl}.

In order to obtain a more accurate value one should also correct for the variations in relative mobilities with composition. Also the question is left, how valid is the assumption of constant $c_{Cl^-} = c_o$? In the computer calculation discussed below, changes in activities and mobilities are corrected for, while no assumption of constant c_{Cl^-} is made.

Computer calculation of concentration profiles in the liquid junction. Diffusion in a liquid junction will lead to changes in concentration profiles with time. The transports by pure diffusion ($j = 0$) for the present HCl–KCl system are given by flux equations (cf. eqs. (4.84):

$$J_{HCl} = -l_{11}\nabla\mu_{HCl} - l_{12}\nabla\mu_{KCl} \tag{8.11a}$$

$$J_{KCl} = -l_{21}\nabla\mu_{HCl} - l_{22}\nabla\mu_{KCl} \tag{8.11b}$$

The changes in local concentrations with time are given by equations similar to eq. (4.73):

$$\partial c_{HCl}/\partial t = -\partial J_{HCl}/\partial x \tag{8.12a}$$

$$\partial c_{KCl}/\partial t = -\partial J_{KCl}/\partial x \tag{8.12b}$$

When the starting conditions are known, the concentration profiles at any time, t, can be calculated from the diffusion coefficients, l_{ij} and the chemical potentials, $\nabla\mu_i$. Then both l_{ij} and $\nabla\mu_i$ must be known as functions of concentration.

Lindeberg[3] carried out computer calculations of concentration profiles at different times from given starting conditions. In order to find suitable expressions for J_{HCl} and J_{KCl} as functions of concentrations and concentration gradients to be used in eqs. (8.12), he made some simplifying assumptions.

The chemical potentials are given by eq. (8.4) and a similar one for KCl. With no assumption about constant c_{Cl^-} there are three variables, c_{H^+} (or c_{K^+}), $c_{Cl^-} = c_{H^+} + c_{K^+}$ and y_{HCl} (or y_{KCl}), where y_{HCl} is given by Harned's rule. The forces $-\nabla\mu_{HCl}$ and $-\nabla\mu_{KCl}$ can be expressed as total differentials of the equations.

The fluxes J_{HCl} and J_{KCl} can be expressed alternatively by means of phenomenological coefficients (cf. eqs. (4.54)):

$$J_{HCl} = -L_{11}\nabla\mu_{HCl} - L_{12}\nabla\mu_{KCl} - L_{13}\nabla\varphi \tag{8.13a}$$

$$J_{KCl} = -L_{21}\nabla\mu_{HCl} - L_{22}\nabla\mu_{KCl} - L_{23}\nabla\varphi \tag{8.13b}$$

$$j = -L_{31}\nabla\mu_{HCl} - L_{32}\nabla\mu_{KCl} - L_{33}\nabla\varphi \tag{8.13c}$$

The diffusion coefficients of eqs. (8.11) are related to the phenomenological coefficients of eqs.(8.13) (cf. eq. (3.16)):

$$l_{ij} = L_{ij} - L_{i3}L_{3j}/L_{33} \tag{8.14}$$

As was discussed in Section 4.1.4, there are relations between the phenomenological coefficients in addition to the Onsager reciprocal relations (4.58) and (4.59):

$$L_{11} + L_{21} = L_{31} \tag{4.58}$$

$$L_{12} + L_{22} = L_{32} \tag{4.59}$$

These equations are valid when the migration of neutral HCl and KCl pairs is negligible. The Nernst–Planck flux equations are based on the assumption that $L_{12} = L_{21} \approx 0$. Lindeberg adopted the same assumption and eqs. (4.58) and (4.59) were reduced to $L_{11} \approx L_{31}$ and $L_{22} \approx L_{32}$, respectively. For pure diffusion, $j = 0$, an expression for J_{HCl} was developed in the same way as eq. (4.67):

$$J_{HCl} = -L_{33}t_{H^+}t_{K^+}(\nabla\mu_{HCl} - \nabla\mu_{KCl}) - L_{33}t_{H^+}t_{Cl^-}\nabla\mu_{HCl} \quad (8.15)$$

A similar equation is obtained for J_{KCl}.

The coefficient L_{33} and the transference numbers can be expressed by concentrations (mole fractions) and mobilities (see eq. (8.3) and cf. eqs. (4.61)–(4.65)). For accurate calculations the variation of mobilities with composition must be included. Lindeberg developed empirical equations for the variation.

For the computer calculation of concentration profiles the liquid junction was taken as a cylinder of 1 cm length between two large reservoirs, HCl solution, 0.1 kmol m^{-3} in electrolyte I, and KCl solution, 0.1 kmol m^{-3} in electrolyte II. The following starting conditions were used:

(1) The cylinder was filled with KCl solution (0.1 kmol m^{-3}) with a sharp boundary facing electrolyte I at the lefthand side.
(2) The cylinder was filled with HCl solution (0.1 kmol m^{-3}) with a sharp boundary facing electrolyte II at the righthand side.
(3) Concentration changes linearly over the cylinder from pure HCl solution at the lefthand side to pure KCl solution at the righthand side.

The type of liquid junction used for cases (1) and (2) is called a *free diffusion junction*. The free diffusion lasts until some diffusing H^+ has reached the right-hand side in case (1), or until diffusing K^+ has reached the left-hand side in case (2). The type used in case (3) is called a *continuous mixture junction*. Since the electrolytes on both sides of the cylinder have fixed composition, the diffusion process must finally come to the same stationary state for all three junctions. Such a junction is called a *restricted diffusion junction*.

The calculated changes in concentration profiles with time for the three starting conditions are shown in Figs. 8.1–8.3.

It can be seen from the figures that the concentration of Cl^- is not constant. It was assumed constant in the liquid junction when using the approximate method for calculating emf.

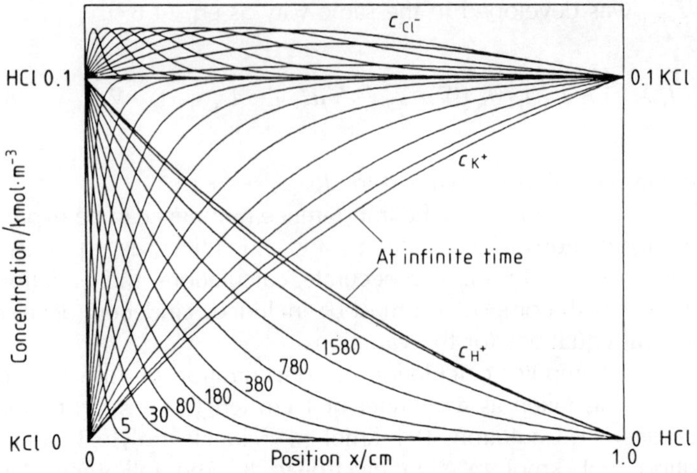

Figure 8.1. Concentration profiles in the liquid junction between an HCl solution of 0.1 kmol m^{-3} (left) and a KCl solution of 0.1 kmol m^{-3} (right). Starting condition: cylinder filled with KCl solution. Curves at times: t/s = 5, 30, 80, 180, 380, 780, 1580, 6380, 12,780 and ∞.

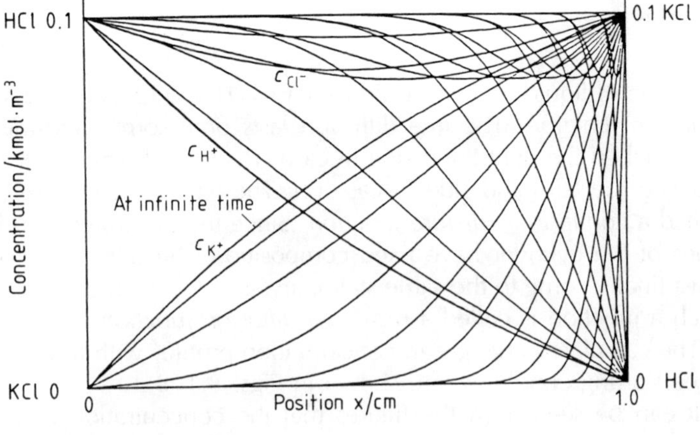

Figure 8.2. Concentration profiles in the liquid junction between an HCl solution of 0.1 kmol m^{-3} (left) and a KCl solution of 0.1 kmol m^{-3} (right). Starting condition: cylinder filled with HCl solution. Curves at times: t/s = 5, 30, 80, 180, 380, 780, 1580, 6380, 12,780 and ∞.

With starting condition (1) (Fig. 8.1) the concentration of Cl^- in the cylinder increases to a higher level because of the higher migration rate for H^+ than for K^+. The high level of concentration spreads from left to right through the cylinder as H^+ migrates to the right. After some H^+ ions have reached the righthand end of the cylinder, the slope of the curve for $c_{H^+} = f(x)$ is no longer equal to zero at $x = 1$. Then the concentration of Cl^- no longer attains as high a level as earlier (after about 1580 s). With starting condition (2) (Fig. 8.2) the concentration of Cl^- in the cylinder decreases for the same reason, i.e. a higher migration rate for H^+ than for K^+. With starting condition (3) (Fig. 8.3) the changes in concentrations with time are small.

As expected, the concentration profiles are the same in all three cases at infinite time, i.e. the stationary state is independent of starting conditions. At the stationary state the concentration of Cl^- has returned very close to the starting concentration after a

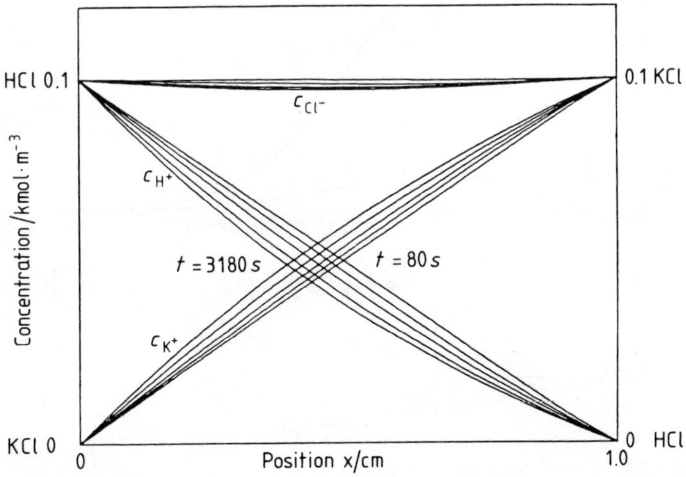

Figure 8.3. Concentration profiles in the liquid junction between an HCl solution of 0.1 kmol m^{-3} (left) and a KCl solution of 0.1 kmol m^{-3} (right). Starting condition: linear concentration changes in cylinder. Curves at times: t/s = 80, 380, 780, 1580 and 3180.

temporary increase or decrease. The maximum deviation is −0.1%. Similar calculations for the systems HCl–NaCl and HCl–LiCl give maximum deviations in concentration of Cl^-, −0.05% and −0.01% respectively at infinite time.

These small deviations in concentration of Cl^- show that the approximation, constant concentration of Cl^-, used when integrating eq. (8.2) to obtain eq. (8.8) is very reasonable when dealing with the stationary state.

When concentration profiles are known, one can calculate the contribution to the emf from the liquid junction at any time. Lindeberg used a numerical method to find this emf as a function of time. The calculation results are shown in Fig. 8.4. The square root of time is used as the independent variable instead of time itself. This simplifies the calculations and the changes come out more clearly in the graph.

We can see from Fig. 8.4 that the emf of the cell may differ by several millivolts, depending on the shape of the concentration

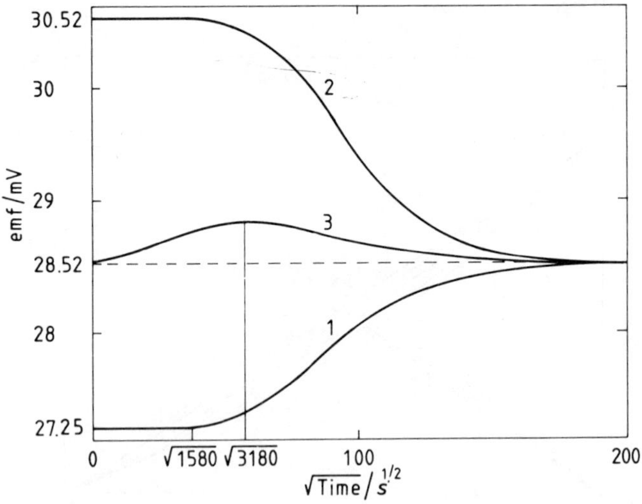

Figure 8.4. The emf as a function of square root of time for the three starting conditions. (1) Free diffusion junction; cylinder contains KCl. (2) Free diffusion junction; cylinder contains HCl. (3) Continuous mixture junction. Concentrations change linearly in cylinder.

profiles. Further, the emf is constant as long as the junction can be characterized as a free diffusion junction, for curve 1 until about 1580 s, when some H^+ ions have reached the right-hand end of the cylinder. After that the emf changes with time. For the continuous mixture junction, curve 3, the emf at the starting point is practically equal to the emf of the restricted diffusion junction obtained at infinite time for all three starting conditions. (The difference between the two emfs is estimated to 0.008 mV).

Experimentally determined values of emf for the cell. For comparison Lindeberg also studied cell a experimentally. The liquid junction consisted of a large number of successive chambers separated by porous glass discs. The chambers were filled with HCl–KCl solutions of constant concentration of Cl^- ($c_{Cl^-} = c_o$) ranging from pure HCl solution on the left-hand side to pure KCl solution on the right-hand side. In this way continuous concentration profiles were imitated by changing the ratio c_{H^+}/c_{K^+} in small steps.

Imitating a continuous mixture junction Lindeberg measured the emf for cell a, as well as for similar cells with HCl–NaCl electrolytes and with HCl–LiCl electrolytes. The experimental values are compared to the computer-calculated values in Table 1.

Table 8.1. The emf of a cell of type a with a continuous mixture junction. Concentration is $c_{Cl^-} = c_o = 0.1$ kmol m^{-3}.

	emf/mV	
System	Experimental	Calculated
HCl–KCl	28.6	28.52
HCl–NaCl	32.8	33.04
HCl–LiCl	35.5	35.44

The good agreement between experimental and calculated values shows that the assumptions underlying the Nernst–Planck flux equations are valid at this concentration. Any coupling between J_1 and X_2 (and between J_2 and X_1) can be neglected, i.e. $L_{12} = L_{21} = 0$. Any migration of neutral ion pairs can also be neglected, i.e. $L_{11} + L_{21} = L_{31}$ and $L_{12} + L_{22} = L_{32}$.

We saw that $c_{Cl^-} \approx c_o$ for the continuous mixture junction as well as for the restricted diffusion junction in the stationary state, and that the emf is the same in both cases (Fig. 8.4). This means that the assumption $c_{Cl^-} = c_o$ made in the approximate calculation is valid when calculating the emf in the stationary state. The result, $E = 26.9$ mV, (eq. (8.9)) is not in the best agreement with the experimental result given in Table 8.1 for the HCl–KCl mixture. The main reason for the discrepancy must be that changes in activity coefficients and in mobilities were neglected.

Other methods of calculating the contribution to the emf from the liquid junction. When the concentrations in electrolytes I and II are different, and when more than three different kinds of ions are present, we can no more use an assumption of a constant concentration throughout the membrane to facilitate the calculation of emf. Methods based on other assumptions were developed by Henderson[4] and by Planck.[5] A discussion of both methods is given by Kotyk and Janáček.[6]

Henderson introduced the assumption of constant concentration gradients (continuous mixture junction). We have seen that the emf in the stationary state is the same as for the continuous mixture junction for the HCl–KCl 0.1 kmol m^{-3} system, and it is a reasonable guess that this will also be the case for more complicated systems. Henderson's method is based on Nernst–Planck flux equations, and for our simple system the method is equivalent to the approximate method leading to eq. (8.8). Henderson's method, however, is not limited to only three kinds of ions. We shall use Henderson's method when studying the salt bridge in Section 8.1.2, and further in Section 8.2.1.

Planck's method does not lead to an explicit expression for emf, but this can be obtained by iterative methods.[7] A simplified, but less accurate, modification of Planck's method was developed by Goldman,[8] who introduced the assumption of constant gradients in electric potential over the junction. For further discussions on liquid junctions and observed emfs for concentration cells see Østerberg and coworkers[9] and Newman.[10]

All the methods above are based on Nernst–Planck flux equations. We have seen that these are valid at concentrations of 0.1 kmol m^{-3}, but how valid are they at higher concentrations? At higher concentrations we cannot neglect coupling, the cross coefficients $L_{12} = L_{21} \neq 0$. Neither can we neglect the diffusion of neutral ion pairs, $L_{11} + L_{21} \neq L_{31}$ and $L_{12} + L_{22} \neq L_{32}$. Miller[11] determined the phenomenological coefficients for mixtures of alkali chlorides in higher concentration ranges (0.45–3 kmol m^{-3}) from experimental values of conductivities, transference numbers and Fick's diffusion coefficients. Miller used forces different from the ones used by Lindeberg, and thus the coefficients were also different. Ratkje[12] gives the relations between Miller's coefficients and Lindeberg's coefficients.

Miller's results give small negative values for L_{12}, -4 to -8% of L_{11}, and positive values for $L_{11} - L_{13}$, $+18$ to $+31\%$ of L_{11}. This means that the use of Nernst–Planck flux equations may give significant errors in emf calculations at higher concentrations.

8.1.2 *The salt bridge*

We have seen that the liquid junction contributes to the emf of an electrochemical cell. A *salt bridge* with a high concentration of KCl interposed between the two electrolytes will reduce the liquid junction potential substantially. Cells where the liquid junction potential has been practically eliminated by a salt bridge are commonly used. One example is the *calomel electrode*, which is used as a constant reference electrode. In this electrode a saturated solution of KCl is used as the salt bridge.

The high concentration of KCl in the salt bridge, and therefore the high conductivity, in addition to the near equal mobilities of K^+ and Cl^- are the main reasons for a low contribution to the emf from a liquid junction to a salt bridge.

Two types of cells with liquid junctions will be considered, represented by cell b and cell c below,

$$\begin{aligned}&(\mathrm{Pt})\mathrm{H_2}(\mathrm{g},1\ \mathrm{atm})|\mathrm{HCl}(\mathrm{aq},c_1)\|\mathrm{KCl}\ (\mathrm{aq},4\ \mathrm{kmol\ m^{-3}})\|\\ &\qquad \mathrm{HCl}(\mathrm{aq},c_3)|\mathrm{H_2}(\mathrm{g},1\ \mathrm{atm})(\mathrm{Pt})\end{aligned} \tag{8.16}$$

$$\begin{aligned}&(\mathrm{Pt})\mathrm{H_2}(\mathrm{g},1\ \mathrm{atm})|\mathrm{HCl}\ (\mathrm{aq},c_1)\|\mathrm{KCl}(\mathrm{aq},4\ \mathrm{kmol\ m^{-3}})\|\\ &\qquad \mathrm{AgCl(s)}|\mathrm{Ag(s)}\end{aligned} \tag{8.17}$$

In cell b (eq. (8.16)) there are two liquid junctions, one on each side of the salt bridge, while cell c (eq. (8.17)) has only one liquid junction. The right-hand side of cell c is a commonly used reference electrode.

The empirical relation between electric potential and concentrations for cell b is

$$\Delta\varphi_b \approx RT \ln c_3/c_1 \tag{8.18}$$

For cell c the empirical relation is

$$\Delta\varphi_c \approx RT \ln c_1 + \text{constant} \tag{8.19}$$

A rigorous calculation of the concentration gradients and the cell emf would require detailed knowledge of activity coefficients and mobilities at high concentration and would involve a great deal of computer work. Approximate calculations by Førland and Østvold[13] have confirmed eqs. (8.18) and (8.19). We shall use Henderson's method[4] to investigate the liquid junction to a salt bridge.

Before studying cells b and c, we shall consider the simpler cell d,

$$\mathrm{Ag(s)|AgCl(s)|HCl(aq,}c_1\mathrm{)\|KCl(aq,}c_2\mathrm{)|AgCl(s)|\ Ag(s)} \tag{8.20}$$

where $c_2 > c_1$.

In order to calculate the emf of the cell we shall assume the concentration gradients, mobilities of ions and activity coefficients to be constant.

The emf of cell d can be expressed by eq. (8.2). Similarly to cell a we have $t_{HCl} = t_{H^+}$ and $t_{KCl} = t_{K^+}$. Transference numbers for the ions are expressed by concentrations and mobilities (cf. eq. (8.3)):

$$t_{H^+} = c_{H^+}u_{H^+}/\Sigma c_i u_i \tag{8.21a}$$

$$t_{K^+} = c_{K^+}u_{K^+}/\Sigma c_i u_i \tag{8.21b}$$

$$t_{Cl^-} = c_{Cl^-}u_{Cl^-}/\Sigma c_i u_i \tag{8.21c}$$

The chemical potential of HCl is given by eq. (8.4) and similarly for KCl. At constant temperature and with constant activity coefficient, the derivatives are

$$d\mu_{HCl} = RT(dc_{H^+}/c_{H^+} + dc_{Cl^-}/c_{Cl^-}) \tag{8.22}$$

and

$$d\mu_{KCl} = RT(dc_{K^+}/c_{K^+} + dc_{Cl^-}/c_{Cl^-}) \tag{8.23}$$

Substituting the expressions for t_i and $d\mu_i$ into eq. (8.2) and using the relation $t_{H^+} + t_{K^+} + t_{Cl^-} = 1$, we obtain

$$\begin{aligned}\Delta\varphi_d &= -RT\int_{(I)}^{(II)} \{(u_{H^+}/\Sigma c_i u_i)dc_{H^+} + (u_{K^+}/\Sigma c_i u_i)\, dc_{K^+} \\ &\quad + (1 - t_{Cl^-})\, dc_{Cl^-}/c_{Cl^-}\} \\ &= -RT\int_{(I)}^{(II)} \{([u_{H^+} - u_{Cl^-}]/\Sigma c_i u_i)\, dc_{H^+} \\ &\quad + ([u_{K^+} - u_{Cl^-}]/\Sigma c_i u_i)\, dc_{K^+}\} - RT \ln c_2/c_1\end{aligned} \tag{8.24}$$

In order to integrate eq. (8.24) distance in the liquid junction must be substituted for concentrations and the gradient in distance for concentration gradients. With constant concentration gradients we have

$$c_{H^+} = c_1\,(1-x) \tag{8.25a}$$

$$c_{K^+} = c_2 x \tag{8.25b}$$

$$c_{Cl^-} = c_1 + (c_2 - c_1)\,x \tag{8.25c}$$

where the distance through the liquid junction, x, is zero at the border to electrolyte I, and unity at the border to electrolyte II. With these substitutions eq. (8.24) is integrated from $x = 0$ to $x = 1$, and we obtain

$$\Delta\varphi_d = -RT\,\frac{(u_{K^+}-u_{Cl^-})c_2 - (u_{H^+}-u_{Cl^-})c_1}{(u_{K^+}+u_{Cl^-})c_2 - (u_{H^+}+u_{Cl^-})c_1} \ln\frac{(u_{K^+}+u_{Cl^-})c_2}{(u_{H^+}+u_{Cl^-})c_1} - RT \ln c_2/c_1$$

or

$$\Delta\varphi_d = -RT \ln c_2/c_1 + \Delta\varphi_{corr} \tag{8.26}$$

The correction term, $\Delta\varphi_{corr}$, originates from differences in mobilities for the ions. It is the same as the calculated *liquid junction potential*.[4] The numerical value of the factor in front of the logarithm decreases with increasing ratio c_2/c_1, while the logarithm itself increases with increasing ratio.

When changes in activity coefficients are considered, an additional correction term is obtained:

$$\Delta\varphi'_{corr} = -RT\int_{(I)}^{(II)} (t_{HCl}\mathrm{d}\ln y_{HCl} + t_{KCl}\,\mathrm{d}\ln y_{KCl}) \quad (8.27)$$

The emf of the cell is then

$$\Delta\varphi_d = -RT\ln c_2/c_1 + \Delta\varphi_{corr} + \Delta\varphi'_{corr} \quad (8.28)$$

For a concentration of KCl, $c_2 = 4$ kmol m^{-3}, in the salt bridge, the two correction terms were calculated for different values of c_1 as shown in Table 8.2. Empirical equations by Lindeberg and Østvold[14] were used for activity coefficients as functions of composition at 25°C.

Table 8.2 shows only small variations with c_1 for the correction terms in the dilute range. The main variation of $\Delta\varphi_d$ with c_1 is given by the term $RT\ln c_1$.

We shall return to cell c. The emf of this cell is equal to the emf of the following series of two cells, $\Delta\varphi_c = \Delta\varphi_e + \Delta\varphi_d$:

(Pt) H_2 (g,1 atm)|HCl (aq,c_1)|AgCl(s)|Ag(s)
⟵ Cell e ⟶
Ag(s)|AgCl(s)|HCl (aq,c_1||KCl (aq,4 kmol m^{-3})|AgCl(s) | Ag(s)
⟵ Cell d ⟶

(8.29)

The emf of cell e on the left-hand side is

$$\Delta\varphi_e = -RT\ln c_1^{\,2} - RT\ln y_{HCl} - (\mu^{o}_{HCl} - \tfrac{1}{2}\mu^{o}_{H_2} - \mu^{o}_{AgCl} + \mu^{o}_{Ag}) \quad (8.30)$$

Adding $\Delta\varphi_d$ (eq. (8.28) for $c_2 = 4$ kmol m^{-3}) we obtain

Table 8.2. Values of the correction terms $\Delta\varphi_{corr}$ and $\Delta\varphi'_{corr}$ for c_2 = 4 kmol m^{-3} and different values of c_1.

c_1/kmol m^{-3}	$\frac{1}{F}\Delta\varphi_{corr}$/mV	$\frac{1}{F}\Delta\varphi'_{corr}$/mV	$\frac{1}{F}(\Delta\varphi_{corr}+\Delta\varphi'_{corr})$/mV
0.001	3.68	9.93	13.61
0.002	3.41	9.74	13.15
0.01	3.06	8.58	11.64
0.02	3.15	7.52	10.67
0.1	4.75	2.91	7.66
0.2	6.47	−0.03	6.44

$$\Delta\varphi_c = -RT \ln c_1 + A \tag{8.31}$$

where

$$A = -RT \ln 4 - (\mu^o_{HCl} - \tfrac{1}{2}\mu^o_{H_2} - \mu^o_{AgCl} + \mu^o_{Ag}) - \{RT \ln y_{HCl} - \Delta\varphi_{corr} - \Delta\varphi'_{corr}\} \tag{8.32}$$

The last term of eq. (8.32) contains the variables that depend on c_1. The variation in the term is small within the dilute range.

We shall now return to cell b. The emf of this cell is equal to the emf of a series of two cells of type c connected by the AgCl|Ag electrode, $\Delta\varphi_b = \Delta\varphi_{c,c_1} - \Delta\varphi_{c,c_3}$. Using eq. (8.31) for each cell c (substituting c_3 for c_1 in the right-hand-side one), we obtain

$$\Delta\varphi_b = RT \ln c_3/c_1 + (A_{c_1} - A_{c_3}) \tag{8.33}$$

The constant terms of A_{c_1} and A_{c_3} cancel and we are left with the variables depending on concentration.

To illustrate the magnitude of the correction $(A_{c_1} - A_{c_3})$ we shall choose an example where c_1 = 0.01 kmol m^{-3} and c_3 = 0.1 kmol m^{-3}. Values for y_{HCl}, $(0.905)^2$ for 0.01 kmol m^{-3} and $(0.796)^2$ for 0.1 kmol m^{-3} are obtained from Harned and Owen,[2] while values for $(\Delta\varphi_{corr} + \Delta\varphi'_{corr})/F$ are found in Table 8.2. The contributions from the two liquid junctions to the emf will partly cancel, and we find

$$E = \Delta\varphi_b/F = \frac{RT}{F} \ln c_3/c_1 - 2.61 \text{ mV} = (59.15 - 2.61) \text{ mV} \tag{8.34}$$

This represents about 4% correction to the approximate formula for $\Delta\varphi_b$ given in eq. (8.18).

For cells with a liquid junction between dilute solutions the emf is practically the same for a continuous mixture junction with linear concentration gradients as for a restricted diffusion junction in the stationary state (see Fig. 8.4). It may be assumed that the difference will also be small for cell b. By computer calculations Lindeberg[3] found a difference of 0.29 mV between the two types of junction.

A more serious uncertainty in the calculations above arises from the use of Nernst–Planck flux equations. As was pointed out in Section 8.1.1, there is substantial coupling of J_{HCl} and J_{KCl} at higher concentrations and substantial diffusion of neutral ion pairs. Empirical emfs obey eqs. (8.18) and (8.19) and we may assume that the deviations from the Nernst–Planck flux equations are not much influenced by variations of concentrations in the dilute electrolyte bordering the liquid junction to the salt bridge. For cell b deviations on the two sides of the salt bridge cancel to a high degree and for cell c the deviation is near constant.

8.1.3 *Conclusions*

A simple liquid junction between two dilute electrolytes gives a substantial contribution, about 30 mV, to the emf of an electrochemical cell. The magnitude of the contribution depends on the concentration profiles in the liquid junction, and variations of more than 3 mV were found.

A liquid junction between a dilute solution and a salt bridge with a high concentration of KCl contributes far less to the emf of a cell. When the salt bridge is interposed between two dilute electrolytes, the contributions to the emf from the two liquid junctions bordering the salt bridge cancel partly. The calculated net contribution to the emf was about one order of magnitude lower than for the simple liquid junction between two dilute solutions. The use of the approximate Nernst–Planck flux equations when higher concentrations are involved makes the result uncertain. Kim and coworkers[15] report data on diffusion, which can be used to improve the calculations.

8.2 *Ion-selective Electrodes*

Ion-selective electrodes specific to numerous different kinds of ions have been made. Cations and anions, both inorganic and organic, can be determined by means of ion-selective electrodes. The electrodes are handy and their use is simple; they have therefore become an indispensible tool in quantitative experimental work.

The selectivity is not equally good in all ion-selective electrodes; there is room for improvement. Even though it is possible to determine a number in the order of magnitude of a hundred different ions, there is still a need to find ion-selective electrodes for more kinds of ions. The yearly output of about 400 papers on ion-selective electrodes shows the great interest in the field (see Ref. 16–20 for some selected textbooks and a review paper on ion-selective electrodes).

All ion-selective electrodes contain membranes and the selectivity of the electrode is determined by the properties of the membrane. We shall start our treatment of ion-selective electrodes with a study of the part played by a membrane in an electrochemical cell (Section 8.2.1). On this basis we shall then study some different types of ion-selective electrodes (Section 8.2.2).

8.2.1 *Cells with ion exchange membranes*

As an example of an ion exchange membrane we shall consider a cation exchange membrane much more permeable to H^+ ions than to Na^+ ions. The principle is the same, however, for all membranes selective to one specific ion, be it a cation or an anion.

We shall consider the electrochemical cell a:

$$\begin{array}{l} Ag(s)|AgCl(s)|KCl(aq,c_3)\|HCl(aq,c_1)NaCl(aq,c_2)|^C| \\ \qquad\qquad\qquad\quad \mathrm{I} \qquad\qquad\qquad\qquad \mathrm{II} \\ \qquad\qquad\qquad\qquad\qquad\qquad HCl(aq,c_0)\,|AgCl(s)|Ag(s) \\ \qquad\qquad\qquad\qquad\qquad\qquad\qquad\quad \mathrm{III} \end{array} \tag{8.35}$$

The cell has three compartments, I, II and III. The left-hand-side AgCl|Ag electrode and compartment I form the reference electrode and it is separated from the test solution in compartment II by means of a liquid junction. The test solution is separated from solution III of the indicator electrode by a cation exchange membrane, $|^C|$. The indicator electrode consists of the membrane, compartment III and the right-hand-side AgCl|Ag electrode. When the membrane is much more permeable to H^+ than to other cations, the cell can be used to obtain the concentration of H^+ in the test solution from experimental values of the emf of the cell.

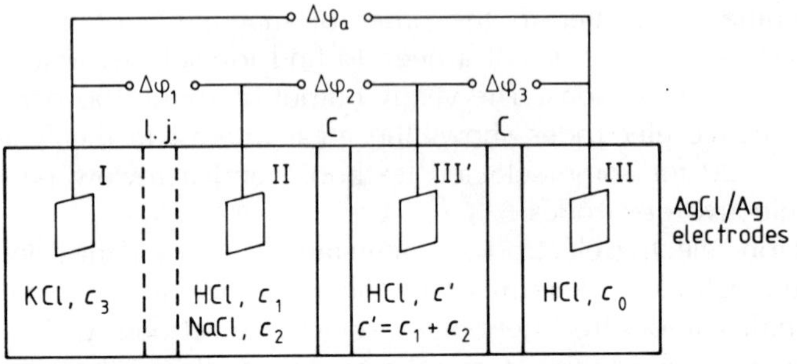

Figure 8.5. A schematic picture of a cell with the same emf, $\Delta\varphi_a$, as cell a. All electrodes are AgCl|Ag electrodes, l.j. is the liquid junction, and C is a cation exchange membrane.

We shall derive the relation between the emf of the cell and the concentrations. For this purpose the cell can be divided into a series of subcells (see Fig. 8.5). To facilitate the calculations an extra compartment III′ is added, separated from II and III by two identical cation exchange membranes. The compartment contains HCl(aq) with a concentration $c' = c_1 + c_2$. Processes at the two electrolyte–membrane interfaces will cancel and the very small contribution to the emf from transport of water is neglected. Hence the addition of compartment III′ does not alter the emf of the cell, and we have

$$\Delta\varphi_a = \Delta\varphi_1 + \Delta\varphi_2 + \Delta\varphi_3 \tag{8.36}$$

where $\Delta\varphi_a$ is the emf of cell a (eq. (8.35)) and $\Delta\varphi_1$, $\Delta\varphi_2$ and $\Delta\varphi_3$ are the emfs of the subcells as shown in Fig. 8.5.

The subcells of Fig. 8.5 resemble cells that were treated earlier and we shall utilize the results of earlier treatments. The calculation of $\Delta\varphi_1$ is analogous to the calculation of emf for cell d (eq. (8.20) of Section 8.1.2). The subcell in the middle with the emf $\Delta\varphi_2$ can be treated in the same way as the cell discussed in Section 5.5, while the right-hand side subcell with emf $\Delta\varphi_3$ is quite similar to the cell discussed in Section 5.3. We shall start with the simplest one, the calculation of $\Delta\varphi_3$.

Calculation of $\Delta\varphi_3$. With the membrane as frame of reference, the change in emf is given by eq. (5.7), $d\varphi = - t_{HCl} d\mu_{HCl} - t_{H_2O} d\mu_{H_2O}$, where $d\varphi$, $d\mu_{HCl}$ and $d\mu_{H_2O}$ anywhere in the membrane are defined as pictured in Fig. 5.4. The emf $\Delta\varphi_3$ is found as the integral of eq. (5.7) over the membrane separating compartments III′ and III. With electrodes reversible to Cl^- and a perfect cation-selective electrode we have $t_{HCl} = t_{H^+} = 1$. In Section 5.3.1 it was found that the integral of the term $t_{H_2O} d\mu_{H_2O}$ gives a very small contribution to the emf. When we neglect this we obtain

$$\Delta\varphi_3 = - \int_{(III')}^{(III)} t_{H^+} d\mu_{HCl} = - \{\mu_{HCl}(c_o) - \mu_{HCl}(c')\} \tag{8.37}$$

Assuming constant activity coefficients we obtain

$$\Delta\varphi_3 = RT \ln (c'/c_o)^2 \tag{8.38}$$

Calculation of $\Delta\varphi_2$. With the membrane as frame of reference, the change in electric potential is

$$d\varphi = - t_{HCl} d\mu_{HCl} - t_{NaCl} d\mu_{NaCl} - t_{H_2O} d\mu_{H_2O} \tag{8.39}$$

This equation can be deduced from eq. (5.58) for $j = 0$. Changes in potentials inside the membrane are defined by differences in potentials between solutions in equilibrium with the membrane as pictured in Fig. 5.8. Equilibrium across the phase boundary membrane-solution is discussed in Section 5.2. Equations (5.52) and (5.53) show how the membrane in the sodium form, NaM, is eliminated, so that the membrane in the hydrogen form, HM, is the frame of reference.

Again we can assume that the integral of $t_{H_2O} d\mu_{H_2O}$ gives only a negligible contribution to the emf of the cell. With electrodes reversible to Cl^- we have $t_{HCl} = t_{H^+}$ and $t_{NaCl} = t_{Na^+}$. The transference numbers are expressed by mobilities in the membrane, u'_{H^+} and u'_{Na^+}, and the mole fractions in the membrane, x_{HM} and x_{NaM}:

$$t_{H^+} = u'_{H^+}x_{HM}/\Sigma u'_i x_{iM} \text{ and } t_{Na^+} = u'_{Na^+}x_{NaM}/\Sigma u'_i x_{iM} \quad (8.40)$$

where $\Sigma u'_i x_{iM} = u'_{H^+}x_{HM} + u'_{Na^+}x_{NaM}$. In the electrolyte we have a constant concentration of Cl^- ions. Assuming constant activity coefficients we have

$$d\mu_{HCl} = RT dc_{H^+}/c_{H^+} \text{ and } d\mu_{NaCl} = RT dc_{Na^+}/c_{Na^+} \quad (8.41)$$

Mole fractions in membrane and concentrations in solution are related by the equilibrium $Na^+ + HM \rightleftarrows H^+ + NaM$. If we have constant activity coefficients in the membrane as well as in the solutions we have

$$K = \frac{c_{H^+}x_{NaM}}{c_{Na^+}x_{HM}} \quad (8.42)$$

(cf. eq. (5.2)).

Neglecting the contribution from the migration of water, we shall integrate eq. (8.39) to obtain $\Delta\varphi_2$. The term t_i is replaced by the expressions in eq. (8.40) and $d\mu_i$ by the expressions in eq. (8.41):

$$\Delta\varphi_2 = -RT\int_{(II)}^{(III')} \{(u'_{H^+}x_{HM}/\Sigma u'_i x_{iM})\, dc_{H^+}/c_{H^+} \quad (8.43)$$

$$+ (u'_{Na^+}x_{NaM}/\Sigma u'_i x_{iM})\, dc_{Na^+}/c_{Na^+}\}$$

Since the concentration of Cl^- is constant, we have $dc_{Na^+} = -dc_{H^+}$. The relation given by eq. (8.42) is used to eliminate all the variables but c_{H^+}, and we obtain,

$$\Delta\varphi_2 = -RT\int_{c_{H^+}=c_1}^{c_{H^+}=c'} \frac{(u'_{H^+} - Ku'_{Na^+})dc_{H^+}}{u'_{H^+}c_{H^+} + Ku'_{Na^+}c_{Na^+}}$$

$$= -RT\ln \{u'_{H^+}c'/(u'_{H^+}c_1 + Ku'_{Na^+}c_2)\}$$

or

$$\Delta\varphi_2 = RT\ln (c_1 + (Ku'_{Na^+}/u'_{H^+})\, c_2) - RT\ln c' \quad (8.44)$$

The emf of a cell where the indicator electrode is combined with an AgCl/Ag electrode is given by the sum $\Delta\varphi_2 + \Delta\varphi_3$, the sum of eqs. (8.38) and (8.44):

$$\Delta\varphi_2 + \Delta\varphi_3 = RT\ln\,(c_1 + (Ku'_{Na^+}/u'_{H^+})c_2) + RT\ln c' - 2\,RT\ln c_0 \qquad (8.45)$$

The selectivity of the electrode to H^+ ions is determined by the ratio Ku'_{Na^+}/u'_{H^+}. For very low values of the ratio, the selectivity is high, and the first term on the righthand side of eq. (8.45) reduces to $RT\ln c_1$. With a high selectivity to H^+ ions, an exchange of H^+ and Na^+ by diffusion through the membrane can be neglected. When there is a high resistance in the circuit of the cell, only negligible currents pass through the cell during measurement of emf ($j \approx 0$), and c_o keeps approximately constant. Recalling that $c_1 = c_{H^+}$ in compartment II, while $c' = c_1 + c_2 = c_{Cl^-}$ we obtain

$$\Delta\varphi_2 + \Delta\varphi_3 = RT\ln c_{H^+}c_{Cl^-} + \text{const} \qquad (8.46)$$

By means of this cell we can determine a concentration *product*, not the concentration of H^+. We shall see how we can determine c_{H^+} in the test solution in compartment II when we combine the indicator electrode with a reference electrode containing a salt bridge. Then we can measure $\Delta\varphi_a$, which is a single-valued function of c_{H^+}. To obtain an equation for $\Delta\varphi_a$, we shall first study the emf $\Delta\varphi_1$ of the left-hand-side subcell pictured in Fig. 8.5.

Calculation of $\Delta\varphi_1$. The present cell is different from cell d (eq. (8.20)) of Section 8.1.2 in two respects. The salt bridge is on the left-hand side in the present cell, which leads to the opposite sign for the emf. The dilute solution in the present cell contains two solutes, leading to a more complicated expression for $\Delta\varphi_{corr}$ than the one in eq. (8.26).

We shall again use Henderson's method[4] (constant concentration gradients in junction), and the same approximations as were used in the derivation of eq. (8.26), constant mobilities and constant activities. Thus we obtain

$$\Delta\varphi_1 = RT\ln\,\{c_3/(c_1+c_2)\} + \Delta\varphi_{corr} \qquad (8.47)$$

where

$$\Delta\varphi_{corr} = RT\,\frac{(u_{K^+}-u_{Cl^-})\,c_3 - (u_{H^+}-u_{Cl^-})\,c_1 - (u_{Na^+}-u_{Cl^-})\,c_2}{(u_{K^+}+u_{Cl^-})\,c_3 - (u_{H^+}+u_{Cl^-})\,c_1 - (u_{Na^+}+u_{Cl^-})\,c_2} \times$$

$$\ln \frac{(u_{K^+}+u_{Cl^-})\,c_3}{(u_{H^+}+u_{Cl^-})\,c_1 + (u_{Na^+}+u_{Cl^-})c_2} \qquad (8.48)$$

Here u_i are ionic mobilities in the electrolyte. When $c_2 = 0$, the correction term above becomes identical (with opposite sign) to the correction term of eq. (8.26).

Some values of $\Delta\varphi_{corr}$ (eq. (8.48)) are given in Table 8.3. The correction term is small, and the variation with concentration is small.

Table 8.3. Values of the correction term $\Delta\varphi_{corr}$ of eq. (8.48) for $c_3 = 4$ kmol m^{-3} and different values for c_1 and c_2.

c_1/kmol m^{-3}	c_2/kmol m^{-3}	$\frac{1}{F}\Delta\varphi_{corr}$/mV
0.01	0.01	−4.24
0	0.1	−1.51
0.01	0.1	−1.84
0.1	0.1	−4.09
0.1	0.2	−3.55

The emf of cell a (eq. (8.35)) is obtained by adding $\Delta\varphi_1$ (eq. (8.47)) and $\Delta\varphi_2 + \Delta\varphi_3$ (eq. (8.45)). Recalling that $c' = c_1 + c_2$ and assuming that $(Ku'_{Na^+}/u'_{H^+})\,c_2 \ll c_1$, we obtain

$$\Delta\varphi_a = RT\ln c_1 + \Delta\varphi_{corr} + RT\ln c_3 - 2\,RT\ln c_o \qquad (8.49)$$

where the two last terms are constant. The equation is based on constant activity coefficients.

When changes in activity coefficients are considered, additional correction tems are obtained (cf. eq. (8.27) of section 8.1.2):

$$\Delta\varphi_{1,corr} = -\,RT\int_{(I)}^{(II)} (t_{HCl}\mathrm{d}\ln y_{HCl} + t_{KCl}\mathrm{d}\ln y_{KCl} + t_{NaCl}\,\mathrm{d}\ln y_{NaCl}) \qquad (8.50)$$

$$\Delta\varphi_{2,\mathrm{corr}} = -RT\int_{(\mathrm{II})}^{(\mathrm{III}')} (t_{\mathrm{HCl}}\mathrm{dln}\, y_{\mathrm{HCl}} + t_{\mathrm{NaCl}}\mathrm{dln}\, y_{\mathrm{NaCl}}) \tag{8.51}$$

$$\Delta\varphi_{3,\mathrm{corr}} = -RT\int_{(\mathrm{III}')}^{(\mathrm{III})} t_{\mathrm{HCl}}\mathrm{dln}\, y_{\mathrm{HCl}} \tag{8.52}$$

In section 8.1.2. empirical equations were used for activity coefficients as functions of composition in mixtures of KCl and HCl. With a lack of corresponding information about mixtures containing NaCl in addition, we have to resort to less accurate estimates of activity coefficients. We shall use an extended Debye–Hückel equation to express the mean activity coefficient:[21]

$$\log_{10} y_{\pm} = -0.5\, z_i^2\sqrt{I}/(1+\sqrt{I}) + Bc_i \tag{8.53}$$

where $I = \frac{1}{2}\Sigma c_i z_i^2 = c_{\mathrm{Cl}^-}$, since all solutes are chlorides of monovalent cations. In dilute solutions we may neglect the term Bc_i and $\log_{10}y_{\pm}$ is approximately the same for all the three salts in solution, HCl, KCl and NaCl. For solutions of about 0.1 kmol m^{-3} the deviation is in the order of magnitude 10%. Recalling that $y_{\pm} = \sqrt{y}$, we thus have

$$\mathrm{dln}\, y_{\mathrm{HCl}} \approx \mathrm{dln}\, y_{\mathrm{KCl}} \approx \mathrm{dln}\, y_{\mathrm{NaCl}} \approx 2\,\mathrm{dln}\, y_{\pm} \tag{8.54}$$

With electrodes reversible to Cl^-, $t_{\mathrm{HCl}} = t_{\mathrm{H}^+}$, $t_{\mathrm{KCl}} = t_{\mathrm{K}^+}$ and $t_{\mathrm{NaCl}} = t_{\mathrm{Na}^+}$. In the liquid junction between compartments I and II, $t_{\mathrm{H}^+} + t_{\mathrm{K}^+} + t_{\mathrm{Na}^+} = 1 - t_{\mathrm{Cl}^-}$. As an approximation an average value of $t_{\mathrm{Cl}^-} = \frac{1}{2}$ may be introduced over the whole concentration range in the liquid junction. Introducing this in eq. (8.50) and substituting eq. (8.54), we obtain

$$\Delta\varphi_{1,\mathrm{corr}} = -RT\int_{(\mathrm{I})}^{(\mathrm{II})} \mathrm{dln}\, y_{\pm} = -RT\ln y_{\pm,\mathrm{II}} + RT\ln y_{\pm,\mathrm{I}} \tag{8.55}$$

In the cation exchange membrane between compartments II and III′ the only current carriers are H^+ and Na^+, hence $t_{\mathrm{H}^+} + t_{\mathrm{Na}^+} = 1$. Introducing this into eq. (8.51) and substituting eq. (8.54), we obtain

$$\Delta\varphi_{2,\text{corr}} = -RT\int_{(\text{II})}^{(\text{III}')} 2\,\text{d}\ln y_{\pm}$$
$$= -2RT\ln y_{\pm,\text{III}'} + 2RT\ln y_{\pm,\text{II}} \qquad (8.56)$$

In the cation exchange membrane between compartments III′ and III the only current carrier is H^+, hence $t_{H^+} = 1$. Introducing this into eq. (8.52) and substituting eq. (8.54), we obtain

$$\Delta\varphi_{3,\text{corr}} = -RT\int_{(\text{III}')}^{(\text{III})} 2\,\text{d}\ln y_{\pm} = -2RT\ln y_{\pm,\text{III}} + 2RT\ln y_{\pm,\text{III}'} \qquad (8.57)$$

The sum of eqs. (8.55), (8.56) and (8.57) is then

$$\Sigma\Delta\varphi_{\text{i,corr}} = +RT\ln y_{\pm,\text{I}} + RT\ln y_{\pm,\text{II}} - 2RT\ln y_{\pm,\text{III}} \qquad (8.58)$$

The first and the third term on the righthand side of eq. (8.58) are constant, independent of the concentrations in the test solution in compartment II. Together with $RT\ln c_3 - 2RT\ln c_o$ of eq. (8.49) they form a constant. The second term can be combined with the first term of eq. (8.49). The emf of cell a with a perfect H^+-selective membrane corrected for the liquid junction potential and for activities is then

$$\Delta\varphi_a = RT\ln c_1 y_{\pm} + \Delta\varphi_{\text{corr}} + \text{const} \qquad (8.59)$$

Neglecting the two corrections, and remembering that $c_1 = c_{HCl} = c_{H^+}$, we obtain the commonly used equation

$$\Delta\varphi_a = RT\ln c_{H^+} + \text{const} \qquad (8.60)$$

As we have seen there are many approximations involved in the derivation of eqs. (8.59) and (8.60). We do not have sufficient data to check the derivation by more detailed calculations. We can, however, check the different assumptions experimentally in a stepwise way, by checking separately the exchange equilibrium across the border membrane–liquid (eq. (8.42) and the three subcells, one by one (see eqs. (8.38), (8.44) and (8.47)).

In deriving $\Delta\varphi_a$ it was seen that the term $RT\ln c_{Cl^-}$ cancelled when adding up the emfs of the subcells. In a similar way the term $RT\ln c_{X^-}$ for any other anion would cancel. We can imagine the series of subcells with an electrode reversible to X^- in the test solution, compartment II. If other anions are substituted for Cl^- and if other cations are added that are not transported in the membrane, the effect on $\Delta\varphi_a$ will only be through the correction term $\Delta\varphi_{corr}$ and the term $RT\ln y_{\pm}$.

The use of single ion quantities. An approach different from the one used above is commonly found in literature on ion-selective electrodes.[16–20] It implies the use of the *electrochemical potential of ions*, $\tilde{\mu}_i$, *single ion chemical potential*, μ_i and *electrostatic potential*, ψ. According to this approach, the main contribution to the emf of cell a comes from the potential difference at the electrolyte–membrane interface, $\Delta\psi_{int}$.

In the treatment below the liquid junction potential will be neglected and we shall limit ourselves to a membrane perfectly selective to H^+ ions.

The local equilibrium at the interface is expressed as

$$\tilde{\mu}_{H^+(aq)} = \tilde{\mu}_{H^+(m)} \tag{8.61}$$

where $\tilde{\mu}_{H^+(aq)}$ is the electrochemical potential of H^+ in the aqueous electrolyte, the test solution containing HCl and NaCl, and $\tilde{\mu}_{H^+(m)}$ is the electrochemical potential of H^+ in the membrane. The electrochemical potential is defined (cf. eq.(4.33)):

$$\tilde{\mu}_{H^+} = \mu_{H^+} + \psi \tag{8.62}$$

and we have

$$\mu_{H^+(aq)} + \psi_{aq} = \mu_{H^+(m)} + \psi_m \tag{8.63}$$

The *interface potential* is equal to the electrostatic potential of the membrane, ψ_m, minus the electrostatic potential of the aqueous electrolyte, ψ_{aq}, and from eq. (8.63) we obtain

$$\Delta\psi_{int} = \psi_m - \psi_{aq} = \mu_{H^+(aq)} - \mu_{H^+(m)} \tag{8.64}$$

In a perfect H^+-selective membrane $\mu_{H^+(m)}$ is constant. We may replace $\mu_{H^+(aq)}$ with $RT\ln a_{H^+}$ + const, and we obtain

$$\Delta\psi_{\text{int}} = RT\ln a_{\text{H}^+} + \text{const} = RT\ln c_{\text{H}^+}y_{\text{H}^+} + \text{const} \qquad (8.65)$$

(Cf. eqs. (8.59) and (8.60)). An approximate calculation of the liquid junction potential may also be carried out with the use of single ion quantities.

None of the quantities $\Delta\psi_{\text{int}}$, a_{H^+} and y_{H^+} are well-defined, measurable quantities. When approximations are used in emf calculations, the use of single ion activities may lead to uncontrolled errors.

8.2.2 *Different kinds of ion-selective electrodes*

Glass electrodes. The oldest and most commonly used ion-selective electrode is the glass electrode for the determination of H^+ concentration. We shall consider the electrochemical cell b consisting of a reference electrode and the glass electrode as the indicator electrode:

$$\text{Ag(s)}|\text{AgCl(s)}|\text{KCl(aq},c_3)\underset{\text{I}}{||}\text{HCl(aq},c_1)\text{NaCl(aq},c_2)\underset{\text{II}}{|}\text{G.m.}|$$

$$\underset{\text{III}}{\text{HCl(aq},c_0)|}\text{AgCl(s)}|\text{Ag(s)} \qquad (8.66)$$

The cell b is very similar to cell a (eq.(8.35)), the only difference being that the cation exchange membrane, $|^{\text{C}}|$, is replaced by a glass membrane, |G.m.|, with some unique properties.

The glass electrode consists of an AgCl|Ag electrode immersed in HCl solution contained in a glass membrane which permits migration of cations. The cation sites of the membrane are non-bridging oxygen ions, Si-0^-. The membrane is mainly a sodium glass, but near the interfaces Na^+ ions have been replaced by H^+ ions from the neighbouring solutions. When preparing a glass electrode for use, it is soaked for some time in an acidic solution to replace Na^+ in the outermost layer with H^+. After treatment the outermost layer is an H^+-containing silica gel of about 100 nm thickness. For a glass with Si-O^- cation sites the following exchange equilibrium:

$$H^+(aq) + \text{Na-glass} \rightleftarrows \text{H-glass} + Na^+(aq) \qquad (8.67)$$

is strongly in favour of H-glass.[22]

Inside the layer of H^+-containing silica gel is a glass zone of increasing Na^+ concentration and decreasing H^+ concentration, about 1000 nm thick. Then there is a core where all the cation sites of the glass are occupied by Na^+ ions, inside which there is again a layer of gradual concentration changes and finally the innermost layer of H^+-containing silica gel. The thickness of the layers depend on glass composition, on the treatment of the glass and on time. Concentration profiles in the glass membrane[20] are shown in Fig. 8.6.

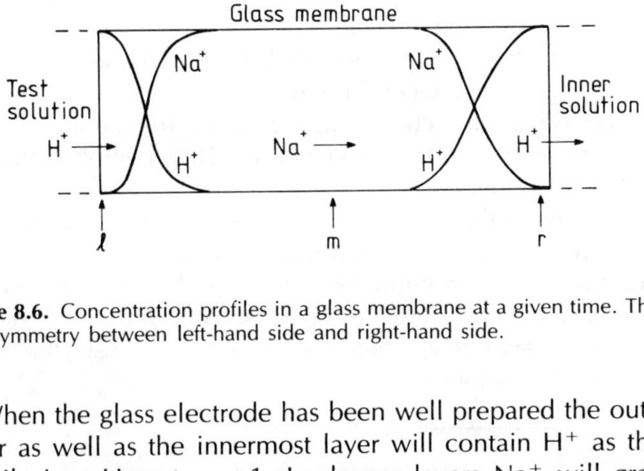

Figure 8.6. Concentration profiles in a glass membrane at a given time. There may be asymmetry between left-hand side and right-hand side.

When the glass electrode has been well prepared the outermost layer as well as the innermost layer will contain H^+ as the only mobile ion. Here $t_{H^+} = 1$. In deeper layers Na^+ will gradually take over as the charge carrier. When the electrode is in use, it may be exposed to solutions with a very low concentration of H^+ and high concentrations of Na^+. The equilibrium (eq. (8.67)), however, is shifted so far to the right that, except for extremely alkaline solutions, we still have $t_{H^+} = 1$ in the outermost layer (and in the innermost layer). The glass membrane is impenetrable to water and the mobile cations are without water of hydration.

We shall derive the relation between the emf of cell b and the concentrations. The cell will be divided into subcells (see Fig. 8.7). The emf of the cell, $\Delta\varphi_b$, is then equal to the sum of the emfs, $\Delta\varphi_1$ and $\Delta\varphi_4$, of the two subcells in series

$$\Delta\varphi_b = \Delta\varphi_1 + \Delta\varphi_4 \tag{8.68}$$

The left-hand-side subcell is equivalent to the left-hand-side subcell for cell a (cf. Fig.8.5) and the emf is given by eq. (8.47):

$$\Delta\varphi_1 = RT\ln\{c_3/(c_1+c_2)\} + \Delta\varphi_{corr} \tag{8.47}$$

We shall study the right-hand-side subcell and the method we shall use to find an expression for $\Delta\varphi_4$ differs from one used in Section 8.2.1 to find $\Delta\varphi_2$ and $\Delta\varphi_3$. In Sections 4.2.1 and 4.2.2 two methods of emf calculation were discussed. The first method is based on the integration of $d\varphi$ obtained from the flux equation for j. In this way eqs. (8.37) and (8.43) were developed. The second method is based on calculating the change in Gibbs energy per faraday transferred in the cell, $\Delta G(Q)$. By the relation $\Delta G(Q) + \Delta\varphi = 0$ (eq. (4.82)) we shall then find $\Delta\varphi$.

An advantage of the second method is that we may consider a cell as consisting of sections and calculate the contribution to $\Delta G(Q)$ from each section separately. The contributions are straightforward additive.

We shall divide the right-hand-side subcell into three sections, compartment II, the glass membrane and compartment III, and calculate the change in Gibbs energy for each one when passing one faraday of electric charge through the cell. Since each section is calculated separately, there is no need to use the same components throughout the cell.

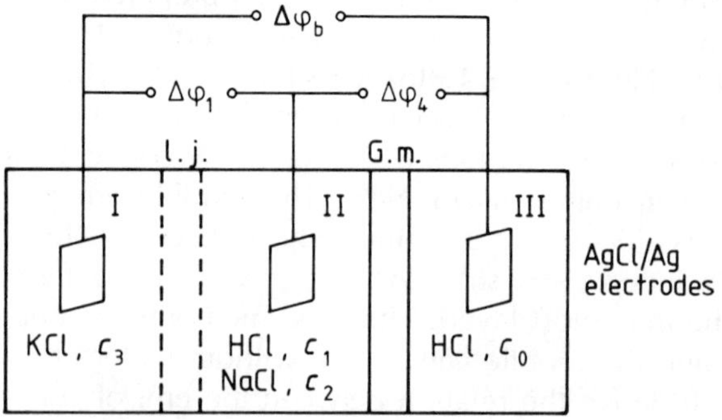

Figure 8.7. A schematic picture of a cell with the same emf, $\Delta\varphi_b$, as cell b. All electrodes are AgCl|Ag electrodes, l.j. is the liquid junction and G.m. is the glass membrane.

The following changes in Gibbs energy occur upon the passage of one faraday of positive charge from left to right through the inner circuit:

Compartment II is an open system with the components HCl, NaCl and H_2O. One mole Cl^- reacts at the electrode, and one mole H^+ enters the glass membrane, $\Delta G_{II} = - \mu_{HCl,II}$.

Glass membrane: see below.

Compartment III is an open system with the components HCl and H_2O. One mole Cl^- is liberated at the electrode, and one mole H^+ is supplied from the glass membrane, $\Delta G_{III} = + \mu_{HCl,III}$.

The *glass membrane* has a constant number of cation sites, and they are always filled with cations to electroneutrality. There must always be an equal number of cations entering and leaving the membrane. Anywhere inside the membrane we must have $dx_{H^+} = - dx_{Na^+}$.

One mole H^+ is supplied to the glass membrane at the interface with compartment II, denoted ℓ in Fig. 8.6. Since $t_{H^+} = 1$ here, all the supplied H^+ migrates to the right and there is no change in Gibbs energy in the outermost layer of the glass. Similarly there is no change in Gibbs energy in the innermost layer facing compartment III at r. Here also $t_{H^+} = 1$, and the supplied H^+ balances with the removed H^+. Within the middle region of the glass, however, there are continuous changes in concentrations of H^+ and Na^+ and in the transference numbers t_{H^+} and t_{Na^+}. The change in Gibbs energy corresponding to these changes is given by the integral

$$\Delta G_{G.m.} = - \int_{(\ell)}^{(r)} (\mu_{HM} dt_{H^+} + \mu_{NaM} dt_{Na^+}) \tag{8.69}$$

where μ_{HM} and μ_{NaM} are chemical potentials of H-glass and Na-glass respectively. By the rules of integration by parts we obtain,

$$\Delta G_{G.m.} = \mu_{HM,\ell} - \mu_{HM,r} + \int_{(\ell)}^{(r)} (t_{H^+} d\mu_{HM} + t_{Na^+} d\mu_{NaM}) \tag{8.70}$$

The electric potential, $\Delta\varphi_{G.m.} = -\Delta G_{G.m.}$, is called an *asymmetry potential*. Khuri[23] discusses the causes of the asymmetry. The two surfaces of the membrane may have different properties because of different treatment during production. Some contribution to the difference may also arise from the difference between $\mu_{HM,\ell}$ and $\mu_{HM,r}$. The integral in eq. (8.70) is equal to zero if the glass membrane behaves as an ideal solution. The integration from ℓ to r may be divided into two steps, from ℓ to m and from m to r, where the limit m is chosen in the core of the membrane where all cation sites are occupied by Na^+. The two integrals will have opposite signs, and, in the ideal case, the same numerical value. When the membrane does not behave ideally, however, asymmetry in the glass membrane may lead to a difference in the numerical values of the two integrals.

The total change in Gibbs energy over the cell is the sum of changes in the three sections:

$$\Delta G(Q) = \Delta G_{II} + \Delta G_{G.m.} + \Delta G_{III} \tag{8.71}$$

The two last terms, $\Delta G_{G.m.}$ and ΔG_{III}, are constant, while ΔG_{II} depends on the composition of the test solution. For an ideal solution in compartment II we have $\Delta G_{II} = -\mu_{HCl,II} = -RT\ln\{c_1(c_1+c_2)\} - \mu^{o}_{HCl}$. Remembering that $\Delta\varphi_4 = -\Delta G(Q)$, we obtain

$$\Delta\varphi_4 = RT\ln\{c_1(c_1+c_2)\} + \text{const} \tag{8.72}$$

Equations (8.47) and (8.72) are added to obtain $\Delta\varphi_b$ for cell b (eq. (8.66)). The concentration of KCl in the salt bridge, c_3, is constant, and we obtain,

$$\Delta\varphi_b = RT\ln c_{H^+} + \text{const} \tag{8.73}$$

where correction terms have been omitted (cf. eq. (8.60)). As for cell a, we may substitute other anions for Cl^-, and any cation that does not enter the glass membrane in competition with H^+ may be present. The glass electrode is an H^+-reversible electrode.

The Si-O$^-$ cation sites in the glass membrane treated above are highly specific for H^+ relative to other cations. Other cation sites in glass, e.g. $[\text{Si-O-Al}]^-$, prefer other cations for H^+

(Eisenman[22]). Glass membranes have been prepared with no Si-O$^-$ cation sites, i.e. all the oxygens are bridged, but with cation sites of the second type. With an appropriate choice of glass composition, the membrane can have a preference for Na^+ ions or for K^+ ions. Thus one can make both Na^+-reversible and K^+-reversible glass electrodes. The specificity of these Na^+ or K^+ electrodes is not so high as for the H^+-reversible glass electrode.

Ionic conducting crystal electrodes. Several ion-selective electrodes based on ionic conducting crystals have been developed. As an example we shall consider the F^- electrode with a membrane consisting of a LaF_3 single crystal. The F^- ions migrate through the LaF_3 crystal lattice.

We shall consider a cell of the same kind as cell a and cell b:

$$\underset{}{Ag(s)|AgCl(s)|}\underset{I}{KCl(aq,c_3)}\|\underset{II}{KF(aq,c_1)KX(aq,c_2)}|LaF_3(s)|$$

$$\underset{III}{NaF(aq,c_0)NaCl(aq,c_4),}|AgCl(s)|Ag(s) \tag{8.74}$$

In this case we shall determine the concentration of the *anion* F^- in compartment II. The cell differs from cell a and cell b in some features. The solutions in compartments I and II have the *cation* in common. When dividing the cell into subcells, we must use an electrode reversible to K^+ in compartment II. In compartment III some Cl^- has been added since the electrode is reversible to Cl^-.

Omitting correction terms, we obtain the emf for the cell:

$$\Delta\varphi = -RT\ln c_{F^-} + \text{const} \tag{8.75}$$

(Cf. eqs. (8.60) and (8.73). The negative sign in front of the logarithmic term in eq. (8.75) is because the anion, F^-, moves in the opposite direction to the electric current. As for the former cells, the presence of other ions in the test solution will influence the correction terms alone, when they cannot be transported through the LaF_3 crystal.

The electrolyte in compartment III can be replaced by a solid electrolyte (Fjeldly and Nagy[24]) to obtain an all-solid-state fluoride electrode of the composition $|LaF_3(s)|AgF(s)|Ag(s)$. The emf of the cell as a function of c_{F^-} in the test solution is given by eq. (8.75).

The constant, however, has a different value.

For other examples of ionic conducting crystal electrodes, see, for example, the textbooks by Cammann[17] and Lakshminarayanaiah[20] and the review paper by Koryta.[18]

Ion-selective electrodes with liquid membrane. The membrane consists of a solution supported by a porous solid matrix of non-conducting material. The solvent is a non-conducting organic liquid. The solute is a salt of the ion to be determined. Both solvent and solute are insoluble in water. Plain inorganic salts are not soluble in organic liquids. In order that the salt be soluble in the organic liquid, both cation and anion must be large, complex organic ions. A simple inorganic ion, such as K^+, can form a complex with an organic carrier to form a soluble salt with a large organic counterion. The electrode is specific to an ion when it migrates readily through the membrane while other ions in the aqueous solutions bordering the membrane do not, and, at the same time, the counterion in the liquid membrane is immobile.

The most well-known example is the K^+-selective liquid membrane electrode, which is widely used in biochemistry for determining potassium in bodily fluids, foods, soils etc. Armstrong and coworkers[25,26] studied the selectivity for K^+ in a liquid membrane consisting of 2,3-dimethylnitrobenzene supported on a cellulose acetate filter. The carrier for the K^+ ion was valinomycin, and the ionogenic group was BPh_4^-, the tetraphenylborate ion. In contrast to other cations, e.g. Na^+, the potassium ion forms a strong complex with the hexadentate ligand valinomycin, hence the content of other ions in the membrane is very low. The ionogenic group is immobile in the membrane, and the valinomycin molecules apparently have a low mobility. When there is an excess of valinomycin over K^+, Armstrong concluded that K^+ ions can exchange rapidly between valinomycin molecules. In this way K^+ carries the current through the membrane.

Specific carriers have been found for other ions, and ion-selective electrodes with liquid membrane have been made for several cations and anions.[17,18,20,27]

8.2.3 *Conclusions*

Most ion-selective electrodes contain a membrane of some kind. The selectivity depends on the equilibria across the interface electrolyte–membrane, on the type of sites for ions in the membrane and on the relative mobilities of competing ions.

We have seen the part played by the salt bridge in a cell for measuring the concentration of an ion. Without a salt bridge between the reference electrode and the indicator electrode, one cannot determine the concentration of an ion separately, but only the activity of a salt.

8.3 *An Energy-storage Battery*

Electrochemical cells with a combination of ion exchange membranes are different from electrochemical cells with liquid junctions. In cells with liquid junctions the chemical reaction takes place at the electrodes, while the combination of membranes permits a reaction in solution in addition to the electrode reactions.

Cells with a combination of membranes are in use for separating dissolved electrolytes by *electrodialysis*. By electrodialysis ions are moved across a membrane from one solution to another under the influence of a direct electric current. The separation of mixed electrolytes by means of electrodialysis requires electric energy. The opposite process, *reverse electrodialysis*, may be used for converting into electric energy, the energy released when mixing electrolytes. One may generate electricity from the mixing of water of different salinity, e.g. where a river reaches the ocean. The mixing of the fresh water of the river with the salt water of the sea dissipates energy corresponding to the energy released by a 200 m waterfall in the river. The harnessing of just a part of this energy by reverse electrodialysis in a salinity power plant may give a substantial energy production.

Electrochemical cells with a combination of membranes can also be utilized for energy-storage batteries. The battery is discharged by the mixing of electrolytes in solution. When this mixing leads to a chemical reaction, the Gibbs energy of the reaction is converted to electric energy. One may recharge the batteries with a direct current forcing the reaction in the opposite direction and demixing of the solutions—or simply by replacing the solutions with new unmixed solutions. It should be possible to develop batteries of this type that are lightweight compared to, for example, the conventional lead storage battery, and which can be recharged very quickly by changing the liquids instead of by a time-consuming electric recharging.

8.3.1 *A cell utilizing acid–base neutralization*

We shall consider the following electrochemical cell:

$$\underset{}{Ag(s)|AgCl(s)|}\underset{I}{KCl(aq)}|^A|\underset{II}{HCl(aq)}|^C|\underset{III}{H_20}|^A|\underset{IV}{KOH(aq)}|^C|\underset{V}{KCl(aq)}|AgCl(s)|Ag(s) \qquad (8.76)$$

The cell has a number of compartments, I, II, III, IV and V, separated by cation exchange membranes $|^C|$ and anion exchange membranes $|^A|$. The electrodes are AgCl|Ag electrodes, and the compartments are filled with electrolytes as shown above.

The cell is an electrochemical energy producer, and Fig. 8.8. shows the directions for the movement of current carriers when an electric current passes from left to right through the inner circuit of the cell. The ions H^+ and OH^- react to form water in compartment III and the cell reaction is

$$HCl(II) + KOH(IV) \rightarrow H_2O(III) + KCl(V) \qquad (8.77)$$

The chemical reaction, neutralization of a strong acid with a strong base, is spontaneous, and the cell should be able to deliver electric work. The conductivity, however, is very low, with pure water in compartment III. In order to improve the conductivity an electrolyte must be added.

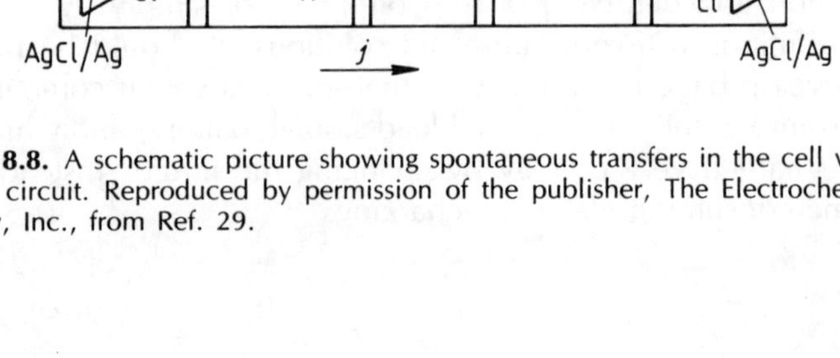

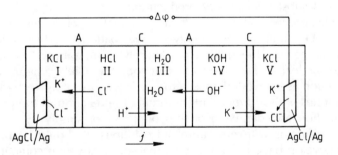

Figure 8.8. A schematic picture showing spontaneous transfers in the cell with a closed circuit. Reproduced by permission of the publisher, The Electrochemical Society, Inc., from Ref. 29.

If we add a salt AX to compartment III, the current is largely carried by A^+ and X^- inside the compartment. When current is passing from left to right in the cell, this will result in a build-up of AOH on the righthand side of the compartment, and of HX on the lefthand side (see Fig. 8.9).

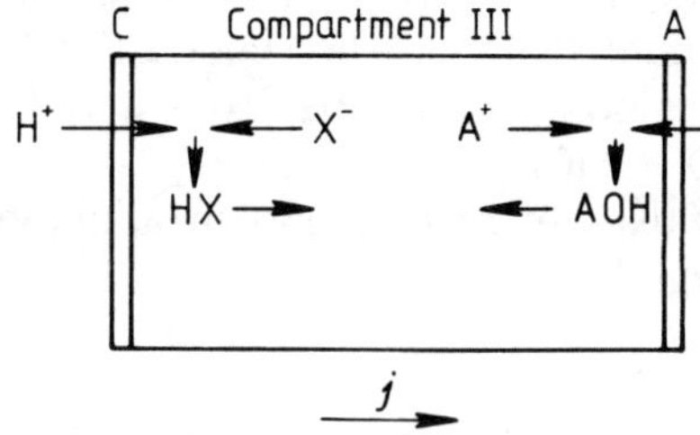

Figure 8.9. Transfers in compartment III when containing a solution of AX.

The HX will diffuse towards the right and AOH towards the left. They will react to form water, X^- and A^+, but this reaction is not coupled to charge transfer. The energy is converted to heat. The result is a loss of electric energy, which will manifest itself as a loss of emf. In order to avoid high losses, one must add an electrolyte which increases the conductivity without leading to a separation of acid and base. We should avoid a variation of pH in the bulk of the solution in compartment III.

When the cell is left with an open circuit, compartment III may become depleted of the electrolyte by the exchange of positive ions with H^+ through the cation membrane on the lefthand side and the exchange of negative ions with OH^- through the anion membrane on the righthand side. To avoid this depletion, the ions of the added electrolyte must be too large to be able to pass through the membranes.

The cell pictured in Fig. 8.8 is a low-voltage cell. In order to increase the voltage one can repeat a unit extending from the middle of compartment I to the middle of compartment V one or several times.

8.3.2 *The theoretical emf of the cell*

The methods for calculating the emf of an electrochemical cell were developed in sections 4.2.1 and 4.2.2.

In order to operate with independent forces only thermodynamic components are used. Local equilibrium is assumed for the cell reaction HCl + KOH = H_2O + KCl anywhere in the cell. (In order to have local equilibrium in compartments II, III and IV, there must be a minute amount of KCl present.) With local equilibrium only three of the four species can be considered as components. We may choose KCl, HCl and H_2O as our set of components. They are given the numbers 1–3. The species KOH, which is not considered as a component, can be expressed by the others, KOH = KCl + H_2O − HCl.

There are three fluxes of matter, and in addition there is the flux of electric charge:

$$J_i = -\sum_{j=1}^{3} L_{ij}\nabla\mu_j - L_{i4}\nabla\varphi; \qquad (i = 1, 2, 3) \tag{8.78a–c}$$

$$j = J_4 = -\sum_{j=1}^{3} L_{4j}\nabla\mu_j - L_{44}\nabla\varphi \tag{8.78d}$$

The corresponding dissipation function is,

$$T\Theta = -\sum_{i=1}^{3} \nabla\mu_i J_i - \nabla\varphi j \tag{8.79}$$

By reasoning similar to that in Section 4.1.1 we find that the coefficient L_{44} is related to the conductivity, $L_{44} = \kappa/F^2$ (cf. eq. (4.6)). We also find that the transference coefficients for the components are given by an equation similar to eq. (4.7):

$$L_{4i}/L_{44} = t_i \tag{8.80}$$

Rearrangement of eq. (8.78d) gives

$$\nabla\varphi = -\sum_{i=1}^{3} (L_{4i}/L_{44})\nabla\mu_i - j/L_{44} \tag{8.81}$$

The electric potential of the cell, $\Delta\varphi$, is obtained by integrating eq. (8.81) for $j = 0$. Using the chemical formulae for the species we obtain

$$\Delta\varphi = -\int_{(I)}^{(V)} (t_{HCl}d\mu_{HCl} + t_{KCl}d\mu_{KCl} + t_{H_2O}d\mu_{H_2O}) \qquad (8.82)$$

(Cf. eqs. (4.68) and (4.71).

In order to integrate this equation one must have knowledge about the transference coefficients for the components. These are related to the transference numbers for the ions (cf. Section 4.1).

When one faraday of electric current passes through the cell from left to right, one mole of Cl^- ions is removed at the left-hand-side electrode and regained at the right-hand-side electrode. The loss of Cl^- in compartment I is compensated for by the transfer of one mole Cl^- ions from compartment II to I, through the anion exchange membrane where $t_{Cl^-} = 1$. To keep the solution in compartment II neutral, one mole H^+ ions must be transferred from compartment II to III, through the cation membrane where $t_{H^+} = 1$. Again this is compensated for by the transfer of one mole OH^- from compartment IV to III, $t_{OH^-} = 1$ through the anion membrane. Furthermore, one mole K^+ must be transferred from compartment IV to V, $t_{K^+} = 1$ throughout the cation membrane. In compartment V the transferred K^+ ions are neutralized by the Cl^- ions regained at the electrode. These transfers are illustrated in Fig. 8.8.

The result of all these changes is that compartment II loses one mole HCl, while compartment III gains one mole H_2O. Simultaneously compartment IV loses one mole KOH and compartment V gains one mole KCl. The species KOH is not a component, and the loss of one mole KOH in compartment IV is equivalent to the loss of one mole KCl and one mole H_2O, and the gain of one mole HCl. Thus the transference coefficient for HCl, t_{HCl}, is unity from compartment II to IV and zero elsewhere in the cell. The transference coefficient for KCl, t_{KCl}, is unity from compartment IV to V and zero elsewhere.

It follows from the definition of the components that the transfer of OH^- from compartment IV to III corresponds to a transfer of H_2O. In addition there will be a transfer of neutral water molecules together with the ions across all membranes. With dilute solutions in all compartments the activity of water is close to unity and the contribution to emf from the transfer of water can be neglected.

With perfect selective membranes and no concentration variation within any compartment, we obtain the following by integration of eq. (8.82):

$$\Delta\varphi = -(\mu_{HCl,IV} - \mu_{HCl,II} + \mu_{KCl,V} - \mu_{KCl,IV}) \tag{8.83}$$

Assuming ideal solutions we can replace chemical potentials with concentration products. When introducing the relation $\mu - \mu^o = RT\ln c_+c_-$ in eq. (8.83), the chemical potentials in the standard state cancel. Furthermore, when the concentration of solute is the same in compartments II, IV and V, we obtain

$$\Delta\varphi = RT(\ln c_{H^+,II} - \ln c_{H^+,IV}) \tag{8.84}$$

We may subtract and add the same quantity, $RT\ln c_{H^+,III}$, to the righthand side of eq.(8.84), and we obtain

$$\Delta\varphi = RT\,\{\ln (c_{H^+,II}/c_{H^+,III}) - \ln (c_{H^+,IV}/c_{H^+,III})\} \tag{8.85}$$

Since $c_{H^+}c_{OH^-} = K_w$, we can replace the term $-\ln(c_{H^+,IV}/c_{H^+,III})$ with $+\ln(c_{OH^-,IV}/c_{OH^-,III})$ and we obtain

$$\Delta\varphi = RT\,\{\ln (c_{H^+,II}/c_{H^+,III}) + \ln (c_{OH^-,IV}/c_{OH^-,III})\} \tag{8.86}$$

The main contribution to the electric potential comes from concentration differences across membranes bordering compartment III. At 25°C the emf of the cell is given by the equation

$$E = -0.059\,\{(pH_{II} - pH_{III}) + (pOH_{IV} - pOH_{III})\} \tag{8.87}$$

For a cell with one unit, pure water in compartment III ($pH = pOH = 7$), $pH = 1$ in compartment II and $pH = 13$ in compartment IV, we obtain $E = 0.71$ V.

8.3.3 *The loss in emf*

The cell containing pure water in compartment III has a very low conductivity. Also, it may be difficult to maintain the high purity of the water. If the membranes enclosing compartment III are not 100% selective, some KOH will diffuse through the anion exchange membrane and some HCl through the cation exchange membrane, contaminating the water in compartment III. On the other hand K^+ will be exchanged with H^+ through the cation exchange membrane and Cl^- with OH^- through the anion exchange membrane. Thus compartment III will contain a solution of mainly KCl (there may be a slight excess of KOH or HCl). The conductivity is improved by the additional charge carriers in compartment III, but even a low concentration of KCl in compartment III leads to a substantial loss in emf.

The interdiffusion of K^+ and H^+ across the cation exchange membrane and of Cl^- and OH^- across the anion exchange membrane are fast processes. In a short time H^+ will have replaced most of the K^+ adjacent to the cation membrane while similarly OH^- will have replaced Cl^- adjacent to the anion membrane. The diffusion in the solution is a much slower process and there will be gradients in concentrations through compartment III. At a given time the momentary concentration profiles in compartment III may be as illustrated in Fig. 8.10.

At the cation exchange membrane on the left-hand side of compartment III, ℓ, all K^+ has been exchanged with H^+, and at the anion exchange membrane on the right-hand side, r, all Cl^- has been exchanged with OH^-, while we have unchanged KCl solution at the neutral point, n, where $c_{H^+} = c_{OH^-}$. In the region ℓ–n we have a mixture of HCl and KCl, while in the region n–r we have a mixture of KCl and KOH. The situation illustrated is early in the diffusion process in a solution originally without concentration gradients. At a later stage the concentration of KCl

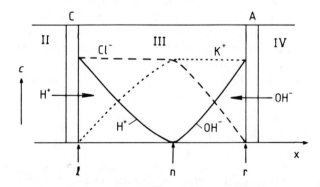

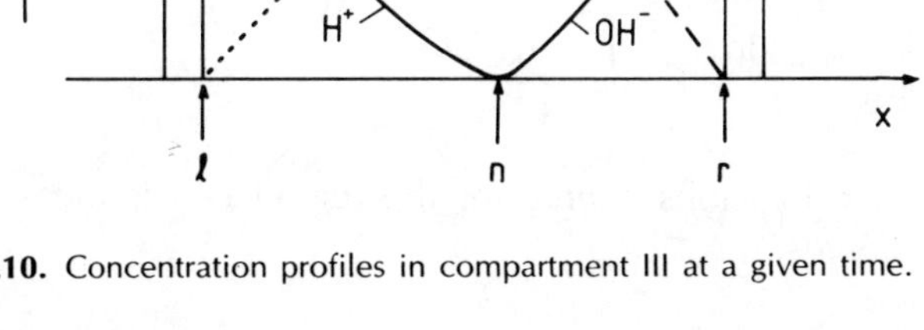

Figure 8.10. Concentration profiles in compartment III at a given time.

at the neutral point will be reduced, maybe to half the concentrations of HCl and KOH at the edges, depending on the relative rates of diffusion in membranes and solution.

Since $\Delta G(Q) + \Delta\varphi = 0$ (see eq. (4.82)), $\Delta\varphi$ for the cell may be found by adding up the changes in Gibbs energy for all the compartments upon the passage of one faraday of electric charge through the cell from left to right. We shall consider the changes in Gibbs energy for a cell containing a solution of KCl in compartment III with a concentration profile as pictured in Fig. 8.10. In all other respects the cell is equal to the one pictured in Fig. 8.8. The following changes in Gibbs energy occur upon the passage of one faraday (compare Section 8.2.2.):

Compartment I: no change, $\Delta G_{\mathrm{I}} = 0$.
Compartment II: one mole HCl is removed, $\Delta G_{\mathrm{II}} = -\ \mu_{\mathrm{HCl,II}}$.
Compartment III: see below.
Compartment IV: one mole KOH is removed, $\Delta G_{\mathrm{IV}} = -\ \mu_{\mathrm{KOH,IV}}$.
Compartment V: one mole KCl is added, $\Delta G_{\mathrm{V}} = +\ \mu_{\mathrm{KCl,V}}$.

In order to find the changes in Gibbs energy for compartment III, it is convenient to consider the two regions ℓ–n and n–r separately.

Compartment III, region ℓ–n: The content of HCl anywhere in this region is equal to the content of H^+. (At the neutral point, n, the content of H^+ is very small, corresponding to the dissociation of pure water, and the content of HCl is equal to zero.) One mole H^+ is supplied at ℓ by migration across the cation exchange membrane. The corresponding change in Gibbs energy is $+\mu_{\mathrm{HCl},\ell}$. Part of the supplied H^+ is removed again by migration to the right, away from ℓ. The corresponding change in Gibbs energy is $-t_{\mathrm{H^+},\ell}\mu_{\mathrm{HCl},\ell}$. Throughout the region ℓ–n there are continuous changes in concentrations of HCl and KCl and in the transference numbers $t_{\mathrm{HCl}} = t_{\mathrm{H^+}}$ and $t_{\mathrm{KCl}} = t_{\mathrm{K^+}}$. The change in Gibbs energy corresponding to these changes is given by the following integrals:

$$-\int_{(\ell)}^{(n)} \mu_{\mathrm{HCl}}\,dt_{\mathrm{H^+}} - \int_{(\ell)}^{(n)} \mu_{\mathrm{KCl}}\,dt_{\mathrm{K^+}}$$

The change in Gibbs energy for the region ℓ–n is the sum of all these changes:

$$\Delta G_{\ell-n} = \mu_{HCl,\ell} - t_{H^+,\ell}\mu_{HCl,\ell} - \int_{(\ell)}^{(n)} \mu_{HCl} dt_{H^+} - \int_{(\ell)}^{(n)} \mu_{KCl} dt_{K^+} \tag{8.88}$$

Remembering that $t_{K^+,\ell}\mu_{KCl,\ell} = 0$ and $t_{H^+,n}\mu_{HCl,n} = 0$, we obtain the following by using the rules of integration by parts:

$$\Delta G_{\ell-n} = \mu_{HCl,\ell} - t_{K^+,n}\mu_{KCl,n} + \int_{(\ell)}^{(n)} t_{H^+} d\mu_{HCl} + \int_{(\ell)}^{(n)} t_{K^+} d\mu_{KCl} \tag{8.89}$$

Compartment III, region n–r: In this region it is convenient to choose KOH, KCl and H_2O as components. By a reasoning similar to that for the region ℓ–n, we can find the change in Gibbs energy by the supply of one mole OH^- through the anion exchange membrane and the migration of ions through the solution. Here $t_{KOH} = -t_{OH^-}$ and $t_{KCl} = -t_{Cl^-}$. We obtain

$$\Delta G_{n-r} = \mu_{KOH,r} - t_{OH^-,r}\mu_{KOH,r} + \int_{(n)}^{(r)} \mu_{KOH} dt_{OH^-} + \int_{(n)}^{(r)} \mu_{KCl} dt_{Cl^-} \tag{8.90}$$

Again using the rules of integration by parts, and since $t_{Cl^-,r}\mu_{KCl,r} = 0$ and $t_{OH^-,n}\mu_{KOH,n} = 0$, we obtain

$$\Delta G_{n-r} = \mu_{KOH,r} - t_{Cl^-,n}\mu_{KCl,n} - \int_{(n)}^{(r)} t_{OH^-} d\mu_{KOH} - \int_{(n)}^{(r)} t_{Cl^-} d\mu_{KCl} \tag{8.91}$$

The total change in Gibbs energy over the cell is the sum of the changes over the compartments:

$$\Delta G(Q) = \Delta G_{II} + \Delta G_{\ell-n} + \Delta G_{n-r} + \Delta G_{IV} + \Delta G_{V} \tag{8.92}$$

We may assume an ideal solution and constant mobilities for the different ions throughout the solution in compartment III. The sum of the four integrals in eqs. (8.89) and (8.91) can then be expressed by the ion mobilities (cf. Section 8.1.1):

$$\int_{(\ell)}^{(n)} t_{H^+} d\mu_{HCl} + \int_{(\ell)}^{(n)} t_{K^+} d\mu_{KCl} - \int_{(n)}^{(r)} t_{OH^-} d\mu_{KOH} - \int_{(n)}^{(r)} t_{Cl^-} d\mu_{KCl}$$

$$= - RT\ln \frac{(u_{K^+}+u_{OH^-})(u_{H^+}+u_{Cl^-})}{(u_{K^+}+u_{Cl^-})^2} \tag{8.93}$$

The sum of the integrals is independent of the concentration of KCl in compartment III. At the neutral point t_{H^+} and t_{OH^-} are negligibly small and therefore $t_{K^+,n} + t_{Cl^-,n} \approx 1$, and thus $t_{K^+,n}\mu_{KCl,n} + t_{Cl^-,n}\mu_{KCl,n} = \mu_{KCl,n}$. We may substitute the expressions for the Gibbs energies on the right-hand side of eq. (8.92) and we may replace $\Delta G(Q)$ with $-\Delta\varphi$. Then we obtain the following expression for the electric potential,

$$\Delta\varphi = - \{\mu_{KCl,V} - \mu_{HCl,II} - \mu_{KOH,IV} + [\mu_{HCl,\ell} - \mu_{KCl,n} + \mu_{KOH,r} - RT\ln \frac{(u_{K^+}+u_{OH^-})(u_{H^+}+u_{Cl^-})}{(u_{K^+}+u_{Cl^-})^2}]\} \tag{8.94}$$

As before we may assume ideal solutions and replace chemical potentials with concentration products in eq. (8.94). Again the chemical potentials in the standard state will cancel. We shall assume the same concentration of solute in compartments II, IV and V. For compartment III we shall assume concentration profiles similar to the ones illustrated in Fig. 8.10, where $c_{HCl,\ell} = c_{KCl,n} = c_{KOH,r}$. We shall also introduce numerical values for ion mobilities. Then we obtain

$$\Delta\varphi = RT\,\{\ln (c_{H^+,II}/c_{H^+,\ell}) + \ln (c_{OH^-,IV}/c_{OH^-,r}) + 1.62\} \tag{8.95}$$

This equation may be compared with eq. (8.86) and it can be seen that the main cause of loss is the separation of H^+ and OH^- in compartment III, or in other words the increase in pH from the left-hand side of the compartment to the right-hand side.

One basis for eq. (8.95) is the assumption of a concentration of KCl at the neutral point unchanged by the diffusion processes (see

Fig. 8.10). If it is reduced by one half of the concentrations of HCl and KOH at left and right respectively, this will give a very small increase in emf. If we replace KCl in compartment III with another electrolyte, AX, we must add the term $(- \mu_{KCl,n} + \mu_{AX,n})$ or for ideal solutions, an equivalent term expressed in concentrations. With ions other than K^+ and Cl^- the last term will be different from 1.62.

In order to obtain a high conductivity with minimum losses in emf, one should add an electrolyte that furnishes many charge carriers, ions, without giving a change in pH through compartment III. As we have seen above, this cannot be accomplished by adding a salt AX that dissociates into a positive and a negative ion. Neither can it be accomplished by adding an acid HX giving the charge carriers H^+ and X^-. When electric current passes from left to right through compartment III, X^- and HX will accumulate on the left-hand side giving an increasing pH from left to right. The addition of a base AOH will have a similar effect on pH.

Some organic substances form *zwitterions*, $^+HY^-$, which can accept a proton to form H_2Y^+, and donate a proton to form Y^- (e.g. amino acids). The ions H_2Y^+ and Y^- will be current carriers. An aqueous solution of a zwitterion will keep a constant pH, the pH of the *isoelectric point*. We may add a zwitterion to the water in compartment III to improve the conductivity of the cell. In order to avoid losses in emf, the number of current carriers added should be so large that the transference numbers of ions from impurities present can be neglected. The number of current carriers at the isoelectric point is determined by the concentration of the zwitterion and the values of the acid constants. The zwitterion should be readily soluble in water. The nearer to each other the values of the acid constants are, the more of the zwitterion reacts with water to form ions, the current carriers. In order to maintain the high conductivity, the zwitterion and its ions should be too large to permeate the membrane.

8.3.4 *Experimental studies of the cell*

The emf of the cell with electrolyte additions to compartment III was investigated experimentally by Makange and coworkers.[28–30]

In a set of experiments known amounts of KCl were added to the water in compartment III, and the corresponding values of emf were observed. The experimental results are shown in Fig. 8.11. A theoretical value for the emf can be calculated from eq. (8.95). We have $\Delta\varphi = nFE$, $c_{H^+,II} = c_{OH^-,IV} = 100$ mol m^{-3} and $c_{H^+,\ell} = c_{OH^-,r} = c_{KCl,III}$. At 25°C this gives

$$E = -\ 0.059 \times 2 \log c_{KCl,III} + 0.28 \quad V \tag{8.96}$$

A fairly good agreement between observed and theoretical values is shown in Fig. 8.11.

The theoretical value of emf for the cell with pure water in compartment III is 0.71 V and thus it can be seen from Fig. 8.11 that losses are substantial when adding KCl to improve the conductivity. Because of membranes that are not perfectly selective, there will be some electrolyte dissolved in compartment III and even without the addition of KCl there are severe losses. When only pure water was placed in compartment III, the observed emf was 0.52 V corresponding to a concentration of electrolyte of

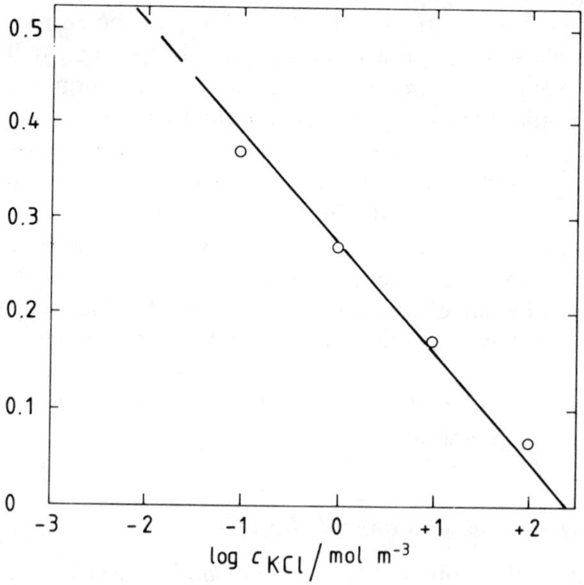

Figure 8.11. The emf as a function of the concentration of KCl in compartment III; ○ = experimental points, line = eq. (8.96). Reproduced by permission of the publisher, The Electrochemical Society, Inc., from Ref. 29.

approximately 10^{-2} mol m^{-3}. The addition of aminonaphthalenedisulphonic acid to a concentration of 10 mol m^{-3} restored the emf to 0.67 V, i.e. 94% of the theoretical voltage.

The above measurements of emf were carried out under the condition $j \approx 0$. It can be seen from eq. (8.78d) that the emf of the cell will decrease with increasing current density. It has been shown experimentally that the decrease in emf with increasing current is much more rapid with pure water in compartment III than when aminonaphthalenedisulphonic acid has been added to the water (see Fig. 8.12). The addition has two effects. First, it increases the conductivity of the water and thus the ohmic loss in electric potential is smaller. Secondly, the solution of the zwitterion has a high buffer capacity, and thus there will be less polarization in compartment III.

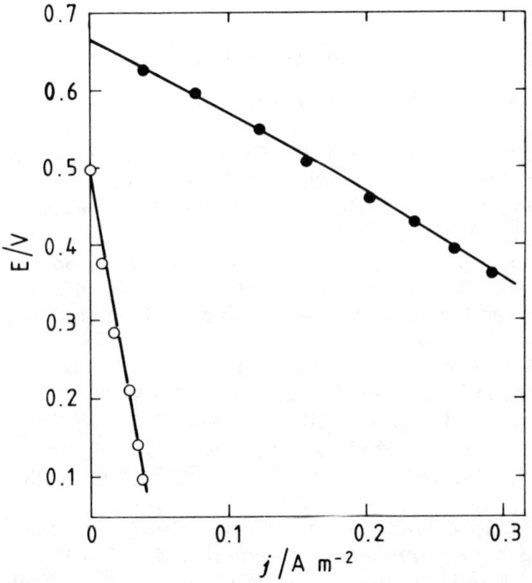

Figure 8.12. The emf of the cell as a function of current density; ○ = pure water in compartment III, ● = aminonaphthalenedisulphonic acid (10 mol m^{-3}) in compartment III.

8.3.5 *Conclusions*

The present cell converts chemical energy into electric energy. The chemical energy is stored as acid and base separated by membranes.

The force of a chemical reaction is a scalar, while transport of electric charge is a vector. According to the Curie principle (Section 3.4) there is no coupling between scalar and vectorial quantities in an isotropic system. In the present cell the membranes make the system anisotropic. The coupling of a vectorial flux and a scalar force implies a vectorial phenomenological coefficient.

Alternatively the scalar force of the chemical reaction can be replaced by a vectorial force. Assuming local equilibrium for the cell reaction, the chemical potential of the species KOH can be expressed by the chemical potentials of HCl, KCl and H_2O. The driving force in the cell is then the difference in chemical potential for HCl, between compartments II and IV—a vectorial force. Vectorial fluxes and forces are coupled by scalar phenomenological coefficients.

8.4 *References*

1. MacInnes, D.A., *The Principles of Electrochemistry*, Reinhold, New York, 1939.
2. Harned, H.S. and Owen, B.B., *The Physical Chemistry of Electrolyte Solutions*, 3rd ed., Reinhold, New York, 1958.
3. Lindeberg, E., *Thesis*, Division of Physical Chemistry, University in Trondheim, N.T.H., 1981.
4. Henderson, P., *Z. Phys. Chem.*, **59**, 118 (1907); **63**, 325 (1908).
5. Planck, M., *Ann. Phys. Chem.*, **39**, 161 (1890); **40**, 561 (1890).
6. Kotyk, A. and Janáček, K., *Biomembranes*, Plenum Press, New York, 1977.
7. Morf, W.E., *Anal. Chem.*, **49**, 810 (1977).
8. Goldman, D.E., *J. Gen. Physiol.*, **27**, 37 (1943).
9. Østerberg, N.O., Jensen, J.B., Sørensen, T.S. and Caspersen, L.D., *Acta Chem. Scand.*, **A34**, 523 (1980).
10. Newman, J.S., *Electrochemical Systems*, Prentice Hall, New York, 1973.
11. Miller, D.G., *J. Phys. Chem.*, **71**, 616 (1967).
12. Ratkje, S.K., *J. Non-Equilib. Thermodyn*, **4**, 75 (1979).
13. Førland, T.and Østvold, T., *The salt bridge in concentration cells, Trans. R. Inst. Technol.*, Stockholm, No. 294, 1972.
14. Lindeberg, E. and Østvold, T., *Acta Chem. Scand.*, **A28**, 563 (1974).
15. Kim, H., Reinfelds, G. and Gosting, L.J., *J. Phys. Chem.*, **77**, 934 (1973).

16. Eisenman, G. (ed.), *Glass Electrodes for Hydrogen and Other Cations*, Marcel Dekker, New York, 1967.
17. Cammann, K., *Das Arbeiten mit ionenselektiven Elektroden*, Springer-Verlag, Berlin, 1973.
18. Koryta, J., *Anal. Chim. Acta*, **183**, 1 (1986).
19. Morf, W.E., *The Principles of Ion-selective Electrodes and of Membrane Transport*, Elsevier, Amsterdam, 1981.
20. Lakshminarayanaiah, N., *Membrane Electrodes*, Academic Press, New York, 1976.
21. Lewis, G.N. and Randall, M., *Thermodynamics*(revised by Pitzer K.S. and Brewer, L.) 2nd ed., McGraw-Hill, New York, 1961.
22. Eisenman, G., The physical basis for the specificity of the glass electrode, in *Glass Electrodes for Hydrogen and other cations* (ed. Eisenman, G.), Marcel Dekker, New York, 1967.
23. Khuri, R.N., Glass microelectrodes and their uses in biological systems, in *Glass Electrodes for Hydrogen and Other Cations* (ed. Eisenman, G.), Marcel Dekker, New York, 1973.
24. Fjeldly, T.A. and Nagy, K., *J. Electrochem. Soc.*, **127**, 1299 (1980).
25. Armstrong, R.D., Covington, A.K., Evans, G.P. and Handyside, T., *Electrochimica Acta*, **29**, 1127 (1984).
26. Armstrong, R.D., Covington, A.K. and Evans, G.P., *Anal. Chim. Acta*, **166**, 103, (1984).
27. Eisenman, G. (ed.), *Membranes*, Vol. 2—*Lipid Bilayers and Antibiotics*, Marcel Dekker, New York, 1973.
28. Makange, A.A., *Thesis*, Division of Physical Chemistry, University in Trondheim, N.T.H., 1980.
29. Førland, K.S., Førland, T., Makange, A.A. and Ratkje, S.K., *J. Electrochem. Soc.*, **130**, 2376 (1983).
30. Makange, A.A., Elektrolytisk celle, *Patent*, 143774 No, (Appl. 772235, F. 24 June 77), 29 Dec 1980.

8.5 *Exercises*

8.1. *Stationary state diffusion.* A porous plug separates two large reservoirs, one containing a solution of the strong electrolyte HX and the other a solution of the strong electrolyte KX. The anion, X^-, is large and practically immobile. The concentration of X^- in solution is assumed to be constant, $c = 0.1$ kmol m^{-3}, throughout the system, reservoirs and plug. The stationary state is obtained in the plug. Transport of water and the small amounts of KX in the HX reservoir and of HX in the KX reservoir are neglected. (a) Derive an equation for J_{H^+} $(= -J_{K^+})$ as a function of the gradient in the mole fraction H^+ in the plug. Assume ideal solutions and use the approximations inherent in the Nernst–Planck equation. Introduce the mobilities u_{H^+} and u_{K^+}. (b) Consider the concentration profiles for H^+ and K^+ in the porous plug. How will they deviate from straight lines when $u_{H^+} > u_{K^+}$? (c) Integrate the equation for J_{H^+} obtained in (a) for a plug of unit area and unit length. Use constant values for the mobilities, $u_{H^+} = 36.3 \times 10^{-8}\ m^2\ s^{-1}\ V^{-1}$ and $u_{K^+} = 7.61 \times 10^{-8}\ m^2\ s^{-1}\ V^{-1}$, and assume that the diffusion takes place only in the x-direction. Neglect the volume occupied by the pore walls in the plug.

8.2. *Calculation of emf.* The following cell is given: Ag(s)|AgCl(s)|HCl(aq,c)||KCl(aq,c)|AgCl(s)|Ag(s). An experimental value of the emf was 28.6 mV at 25°C when c = 0.1 kmol m^{-3}. This value shall be compared to theoretically calculated values. For an approximate calculation we shall assume constant concentration of Cl^- throughout the cell. Use constant mobility ratios, $u_{H^+}/u_{Cl^-} = 4.6$ and $u_{K^+}/u_{Cl^-} = 0.96$. (a) Calculate the emf assuming ideal solutions. (b) Calculate the emf using the following expressions for activity coefficients, $\ln y_{HCl} = \text{const} - 177 \times 10^{-6}\ c_{KCl}$ and $\ln y_{KCl} = \text{const} + 177 \times 10^{-6}\ c_{HCl}$. Compare your result with the calculated value in Table 8.1, obtained by a more accurate method.

CHAPTER 9

Transport Properties of a Cation Exchange Membrane

Membranes are becoming more and more important in chemical technology, and the number of applications to industrial processes is increasing.

Some well-known membrane processes were mentioned in Chapter 5 (introduction). In Section 8.2, ion-selective electrodes, which are an important utilization of membranes, were discussed. For many years membranes have been used to keep solutions separate in electrolysis; Section 8.3 gives an example of how a set of membranes makes it possible to convert the chemical energy of a non-redox reaction to electric energy.

Separation processes represent a common use of membranes. Evaporation processes can be replaced by membrane separation or can be combined with a pretreatment by membrane separation to reduce the cost. Two types of membrane separation processes are *electrodialysis* and *reverse osmosis*.[1,2] In electrodialysis, an electric current is used to transfer salt from a dilute solution to a more concentrated solution through ion exchange membranes. In reverse osmosis pressure is used to force water out of a concentrated solution through a membrane.

The methods are used for desalination of water and for purifying waste water for reuse. Great investments are made in producing water for drinking and for industrial use in this manner. Both methods are also used in food processing. Deacidification and dewatering of fruit juices is one example. The methods are useful in pollution control and waste recovery. A common application is the treatment of whey-bearing dairy waste waters.

In the production of sodium hydroxide and chlorine by electrolysis in aqueous solution the diaphragm is being replaced by a cation exchange membrane. This prevents the brine from passing into the cathode compartment, as anions cannot go through the membrane.

In future we may expect an extended use of membranes by methods already tested and a widening scope for the application of membranes.

Phenomenological coefficients for the membrane are used in the description of electrodialysis, reverse osmosis and several other membrane processes. Central coefficients are electric conductivity, transference coefficients and water permeability. In order to interpret the coefficients we must know the equilibria across the phase boundaries membrane-solution. A theoretical background was given in Chapter 5. In this chapter we shall study the values of phenomenological coefficients for a commercial cation exchange membrane (CR-61 AZL 386 from Ionics Inc., Watertown, MA, US). The membrane will be denoted CR. Experimental data obtained from transport studies of Na^+–H^+ mixtures and of K^+–Sr^{2+} mixtures will be used as illustrations.[3–6] With a background of data we shall discuss the Nernst–Planck equations and other approximations.

We shall not deal with the practical problems involved in the operation of plants using membrane technology, such as swelling, scaling and fouling of membranes. One important problem is concentration polarization, which may influence the fluxes strongly (see Weber[7]).

9.1 *Exchange Equilibrium Membrane–Solution*

Equilibrium across the phase boundary membrane–solution was discussed in Section 5.2. We shall look at the specific equilibria for the cation pairs Na^+, H^+ and K^+, Sr^{2+}.

A cation exchange membrane in contact with an aqueous solution of NaCl and HCl has all cation sites, M^-, filled with Na^+ and H^+. The membrane components are NaM and HM respectively. The exchange equilibrium can be written as

$$\mathrm{HCl(aq)} + \mathrm{NaM} \rightleftarrows \mathrm{NaCl(aq)} + \mathrm{HM} \tag{9.1}$$

A similar exchange equilibrium for the ion pair K^+, Sr^{2+} is

$$\mathrm{SrCl_2(aq)} + 2\ \mathrm{KM} \rightleftarrows 2\ \mathrm{KCl(aq)} + \mathrm{SrM_2} \tag{9.2}$$

Experimental values are obtained for Gibbs energies for the reactions given in eq. (9.1)[6] and eq. (9.2).[8] The thermodynamic

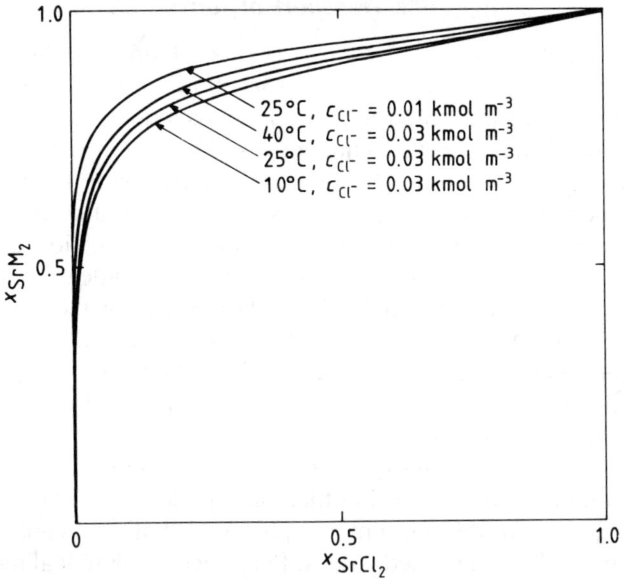

Figure 9.1. Equilibrium isotherms for cation exchange membrane CR. Equivalent fraction of Sr in membrane, x_{SrM_2}, as function of equivalent fraction of Sr in solution, x_{SrCl_2}, for different temperatures and two different total concentrations, c_{Cl^-}, in the solution. From: T. Holt *et al.*, *J. membrane Sci.*, **25** (1985) 133. Copyright © 1985 Elsevier Science Publishers B.V. Reprinted by permission of Elsevier Science Publishers B.V.

equilibrium constants are $K_1 = 1.4$ and $K_2 = 5.6$ at 25°C. Experimental equilibrium isotherms for eq. (9.2)[8] are shown in Fig. 9.1. *Equivalent fractions*, $x_{SrCl_2} = 2\,n_{SrCl_2}/(n_{KCl} + 2\,n_{SrCl_2})$ and $x_{SrM_2} = 2\,n_{SrM_2}/(n_{KM} + 2\,n_{SrM_2})$ are given on the abscissa and the ordinate respectively. The results show the preference for Sr^{2+} over K^+ in the membrane, a result that is typical for membranes containing ions of different valence.[9] From the value of K_1 we can deduce that equilibrium isotherms for eq. (9.1) would show only a slight preference for H^+, the smaller ion, over Na^+ in the membrane.

The water content in ion exchange membranes varies from 30 to 70% by volume, depending on ionogenic groups and activity of water in adjacent solution.[9] The present membrane, CR, contains 40% water.[8] This corresponds to 17 moles of water per monovalent cation site, M^-. Up to a total concentration $c_{Cl^-} = 0.03$ kmol m^{-3}, the content of mobile anions in the membrane is negligible.[8]

9.2 *Transport of Ions*

A concentration cell where two aqueous solutions containing the cation pair Na^+, H^+ are separated by a cation exchange membrane, is treated in Section 5.5. The flux equations (eqs. (5.54)) are given in Section 5.5.1. The flux equations for the cation pair K^+, Sr^{2+} are analogous to these. The present chapter refers to particular phenomenological coefficients with subscripts, and they are defined by eqs. (5.54) or the analogous equations for the cation pair K^+, Sr^{2+}. Diffusion coefficients and transference coefficients refer to the flux equations (eqs. (5.59) in Section 5.5.2) or the analogous equations for the cation pair K^+, Sr^{2+}. In Section 5.5.2 relations between coefficients are derived for the cation pair Na^+, H^+. A generalization of the derivation is given and this also covers the cell with the cation pair K^+, Sr^{2+}.

The principles for the experimental determination of coefficients are given in Section 5.5.3, but there are no details on the methods to be used. In the present chapter we shall concentrate on experimentally determined values. References to practical methods will be given.

9.2.1 *The conductivity of the membrane. Ionic mobilities and ionic transference numbers in the membrane*

The coefficient L_{44} represents the conductivity of the membrane. It can be expressed by the mobilities of the ions in the membrane, u_i, and the equivalent fractions of the ions in the membrane, x_i. For a membrane that is a mixture of NaM and HM we have

$$L_{44} = \frac{c}{F}(x_{NaM}u_{Na^+} + x_{HM}u_{H^+}) \tag{9.3}$$

where c is the concentration of cation sites in the membrane (cf. eq. (4.61) which represents the conductivity of a liquid junction). The conductivity, $\kappa = L_{44}F^2$ (cf. eq. (4.6)). We have similar relations for a membrane that is a mixture of KM and SrM_2.

Since $x_{NaM} + x_{HM} = 1$, x_{NaM} can be eliminated from eq. (9.3), and we obtain

$$\kappa = cF\,\{(u_{H^+} - u_{Na^+})\,x_{HM} + u_{Na^+}\} \tag{9.4}$$

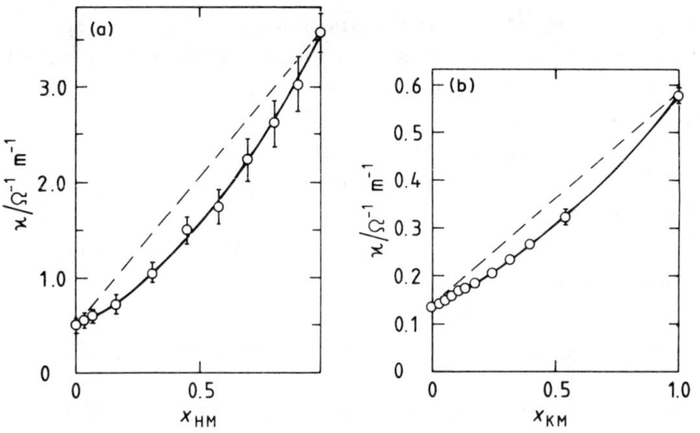

Figure 9.2. Membrane conductivity as a function of equivalent fraction (a) of HM in a membrane consisting of HM and NaM, (b) of KM in a membrane consisting of KM and SrM_2. ⧲ Experimental points with uncertainties. Refer to the text concerning the curved lines.

Mobilities of ions in dilute aqueous solutions are essentially constant, as ions can move independently. If mobilities were constant in the membrane, eq. (9.4) would give κ as a linear function of x_{HM}.

Experimental results for κ as a function of x_{HM} are presented graphically in Fig. 9.2. As can be seen from the figure, a straight line cannot be fitted to the results. This means that u_{Na^+} and u_{H^+} of eq. (9.4) do not have constant values.

An ion exchange membrane is quite different from an aqueous solution. Electroneutrality requires that all cation sites, M^-, in our cation exchange membrane are filled with cations, and one cannot expect independent movements of the cations. On the contrary, one should expect obstructions to movements by collisions.

The following equations for the mobilities are proposed:

$$u_{H^+} = u^o_{H^+}(1-kx_{NaM}) \tag{9.5a}$$

$$u_{Na^+} = u^o_{Na^+}(1-kx_{HM}) \tag{9.5b}$$

where $u^o_{H^+}$ and $u^o_{Na^+}$ are the mobilities of H^+ in a pure HM membrane and of Na^+ in a pure NaM membrane, respectively. The parameter k is a constant chosen to give the best fit of the curve to the experimental points in Fig. 9.2a.

The rationale behind these equations is a reduction in the mobility of an ion proportional to the number of neighbour ions of the other kind. Replacing u_{H^+} and u_{Na^+} with eqs. (9.5a) and (9.5b), respectively, in eq. (9.4) we obtain κ as a second-order polynominal of x_{HM}:

$$\kappa = cF\,\{x_{NaM}u^o_{Na^+}\,(1-kx_{HM}) + x_{HM}u^o_{H^+}(1-kx_{NaM})\} \qquad (9.6)$$

This equation is represented by the curved line in Fig. 9.2a. The line corresponds to a value of the parameter, $k = 0.50$. Experimental values for the conductivities in pure HM membrane and pure NaM membrane are 3.6 $\Omega^{-1}m^{-1}$ and 0.5 $\Omega^{-1}m^{-1}$, respectively. The concentration of cation sites in the membrane is 1.6 k equivalent m^{-3},[3] and the mobilities in pure membranes are obtained, $u^o_{H^+} = 2.3 \times 10^{-8}$ ms^{-1}/Vm^{-1} and $u^o_{Na^+} = 0.3 \times 10^{-8}$ ms^{-1}/Vm^{-1}. The mobility ratio u_{H^+}/u_{Na^+} varies from 4 in an almost pure NaM membrane to 15 in an almost pure HM membrane.

For a membrane containing the cation pair K^+, Sr^{2+} the conductivity can be expressed as a function of x_{KM} by an equation similar to eq. (9.4). Assuming functions for mobilities similar to eq. (9.5), an equation similar to eq. (9.6) is obtained. The best fit of the curve is found for $k = 0.28$. Experimental values for the conductivities in pure KM membrane and pure SrM_2 membrane are 0.58 Ω^{-1} m^{-1} and 0.14 Ω^{-1} m^{-1}, respectively (see Fig. 9.2b). The corresponding mobilities are $u^o_{K^+} = 0.38 \times 10^{-8}$ ms^{-1}/Vm^{-1} and $u^o_{Sr^{2+}} = 0.09 \times 10^{-8}$ ms^{-1}/Vm^{-1}. The ratio $u_{K^+}/u_{Sr^{2+}}$ varies from 3 in an almost pure SrM_2 membrane to 6 in an almost pure KM membrane.

The mobilities of the ions in the membrane are about one order of magnitude lower than the mobilities for the same ions in aqueous solution, as given by Harned and Owen.[10] The order of magnitude of the mobilities in the membrane is typical for commerical membranes.[11,12] The ratio $u^o_{K^+}/u^o_{Na^+}$ is 1.3, compared to the ratio 1.5 for the mobilities of the same ions in an infinitely dilute aqueous solution. The mobility of Sr^{2+} is relatively more decreased than the mobilities of the monovalent ions.

Ionic transference numbers in membrane can be expressed by equivalent fractions and mobilities. In the membrane consisting of HM and NaM we have

$$t_{H^+} = L_{14}/L_{44} = x_{HM}u_{H^+}/(x_{HM}u_{H^+} + x_{NaM}u_{Na^+})$$
$$= x_{HM}u_{H^+}cF/\kappa \qquad (9.7a)$$
$$t_{Na^+} = L_{24}/L_{44} = x_{NaM}u_{Na^+}/(x_{HM}u_{H^+} + x_{NaM}u_{Na^+})$$
$$= x_{NaM}u_{Na^+}cF/\kappa \qquad (9.7b)$$

(Cf., for example, eqs. (4.24), (4.63) and (4.64).) For any composition of the membrane, the values for t_{H^+} and t_{Na^+} can be calculated from eqs. (9.6) and (9.7). It can be seen from Fig. 9.2(a) that experimental values have some uncertainties, larger when there is more HM in the membrane. This gives some uncertainty to values of u_i^o and k and corresponding uncertainties in the values of t_i.

Table 9.1 gives values of t_{H^+} for some membrane compositions calculated from eq. (9.7a). The values are compared to experimental results obtained by the Hittorf method[6] (see Section 4.1.1 for the principles of the Hittorf method). The two sets of transference numbers agree within experimental uncertainties. Recently an improved experimental procedure for the determination of transference numbers in membranes has been reported.[13]

In a similar way t_{K^+} and $t_{Sr^{2+}}$ can be calculated from the results of the conductivity measurements. It can be seen from Fig. 9.2(b)

Table 9.1. Transference numbers for H^+ in cation exchange membrane CR for the system HM–NaM.

Membrane composition x_{HM}	Calculated transference number $t_{H^+} = x_{HM}u_{H^+}cF/\kappa$	Experimental transference number t_{H^+}
0	0	—
0.161	0.48 ± 0.03	0.48 ±0.03
0.452	0.86 ± 0.06	0.78 ± 0.03
0.702	0.96 ± 0.07	0.93 ± 0.03
0.911	0.99 ± 0.07	0.99 ± 0.03
1	1	—

that experimental uncertainties are smaller, so the values will be more precise than for t_{H^+}.

Experimental values for conductivities can give the coefficient L_{44}. With known values for the transference numbers, experimental or calculated, L_{14} and L_{24} can be obtained.

9.2.2 *Validity of Nernst–Planck equations for cation exchange membranes*

The Nernst–Planck flux equations for cation exchange membranes were discussed in Section 5.5.2. Equations (5.65), $J_1 = -L_{14}(\nabla\mu_1+\nabla\varphi)$ and $J_2 = -L_{24}(\nabla\mu_2+\nabla\varphi)$, for a membrane containing only monovalent ions, are approximations, neglecting L_{12} and $\nabla\mu_3$. When water vapour pressure is the same over both electrolytes in contact with the membrane, we may assume that $\nabla\mu_3$ is small, unless diffusion of ions leads to pressure gradients *inside* the membrane. The possibility of pressure variations inside the membrane will be discussed in Section 9.3.1. In this section we shall study the conditions for neglecting L_{12}.

The Nernst–Planck flux equations can be developed from the general flux eqs. (5.54) when the relations between the fluxes of ions and the current density are considered:

$$z_1J_1 + z_2J_2 = j \tag{9.8}$$

where z_i is the charge of the cation. This gives the following relations between coefficients:

$$z_1L_{1i} + z_2L_{2i} = L_{4i}; \qquad (i = 1,2,3,4) \tag{9.9}$$

(Cf. eqs. (5.64) and (5.67)). For the two cases studied here eq. (9.9) gives

$$\text{HM–NaM membrane:} \qquad L_{11} + L_{21} = L_{41} \tag{9.10}$$

$$\text{KM–SrM}_2\text{ membrane:} \qquad L_{11} + 2L_{21} = L_{41} \tag{9.11}$$

When one assumes $L_{21} \approx 0$ and $\nabla\mu_3 \approx 0$, the following Nernst–Planck equations are obtained:

$$\text{HM–NaM membrane:} \quad J_{H^+} = -L_{14}(\nabla\mu_{HM} + \nabla\varphi) \tag{9.12a}$$

$$J_{Na^+} = -L_{24}(\nabla\mu_{NaM} + \nabla\varphi) \tag{9.12b}$$

$$\text{KM–SrM}_2 \text{ membrane:} \quad J_{K^+} = -L_{14}(\nabla\mu_{KM} + \nabla\varphi) \qquad (9.13a)$$

$$J_{Sr^{2+}} = -L_{24}\left(\tfrac{1}{2}\nabla\mu_{SrM_2} + \nabla\varphi\right) \qquad (9.13b)$$

The determination of the coefficients L_{14} and L_{24}, as well as L_{24}, for both systems was discussed in Section 9.2.1. With an independently determined coefficient L_{11}, eqs. (9.10) and (9.11) can be used to check the assumption L_{21} $(= L_{12}) \approx 0$.

The phenomenological coefficient L_{11} is related to the diffusion coefficient l_{11} (cf. eq. (5.60)):

$$L_{11} = l_{11} + L_{14}L_{14}/L_{44} \qquad (9.14)$$

In order to obtain experimental values for l_{11}, fluxes were observed in pure diffusion experiments.

The cell, HCl(aq),NaCl(aq)|membrane CR|HCl(aq), with $c_{Cl^-} = 0.03$ kmol m^{-3} in both aqueous solutions, was used to determine J_{NaCl} as a function of x_{HCl} in the left-hand-side solution.[6] The result is given in Fig. 9.3(a).

In a similar way J_{SrCl_2} was determined as a function of x_{KCl} in the lefthand-side solution of the cell, KCl(aq),$SrCl_2$(aq) |membrane CR|KCl(aq), with $c_{Cl^-} = 0.03$ kmol m^{-3} in both aqueous solutions.[3] The result is given in Fig. 9.3(b).

With the knowledge of concentration profiles in the membrane, the local values of l_{11} can be calculated for the fluxes J_{Na^+} and $J_{Sr^{2+}}$ in the two systems, HM–NaM and KM–SrM_2, respectively. Holt and coworkers[4] discuss how concentration profiles, and hence the local values of l_{11}, are obtained from the above results. The value of L_{11} is found by means of eq. (9.14).

The assumption $L_{12} \approx 0$ is tested for the system HM–NaM by the application of eq. (9.10) to experimental values of L_{11} and L_{41}, and for the system KM–SrM_2 by the application of eq. (9.11). The results are shown graphically in Fig. 9.4.

Figure 9.4 shows that L_{12} is very small compared to L_{11} and L_{14} in both systems. The neglect of L_{12} in the Nernst–Planck equations leads to a very small error only in quantitative values for transports in the membranes. This means that mechanisms of transport are similar for diffusion and conduction. For both systems $L_{11} < L_{14}$, leading to a positive value for L_{12} over the whole range of composition.

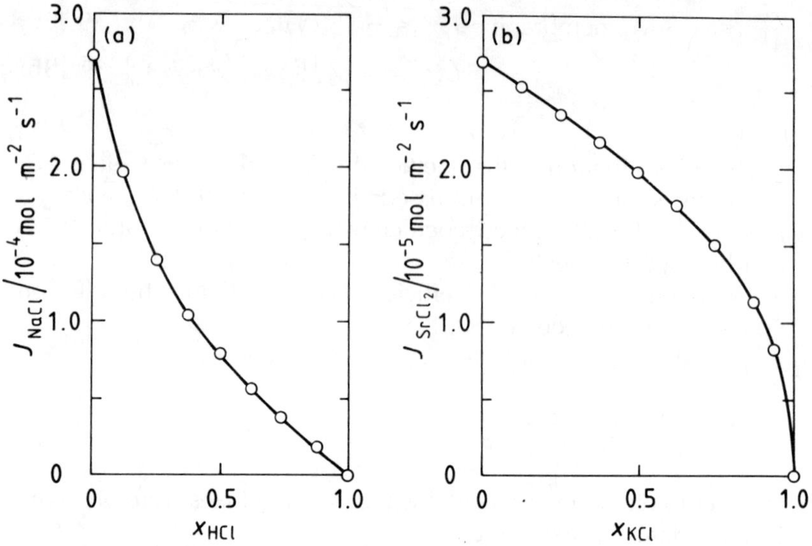

Figure 9.3. Diffusion fluxes as function of composition of electrolyte.
(a) The cell, HCl(aq),NaCl(aq)|membrane CR|HCl(aq) $c_{Cl^-} = 0.3$ kmol m^{-3}.
(b) The cell, KCl(aq),$SrCl_2$(aq)|membrane CR|KCl(aq) $c_{Cl^-} = 0.3$ kmol m^{-3}.

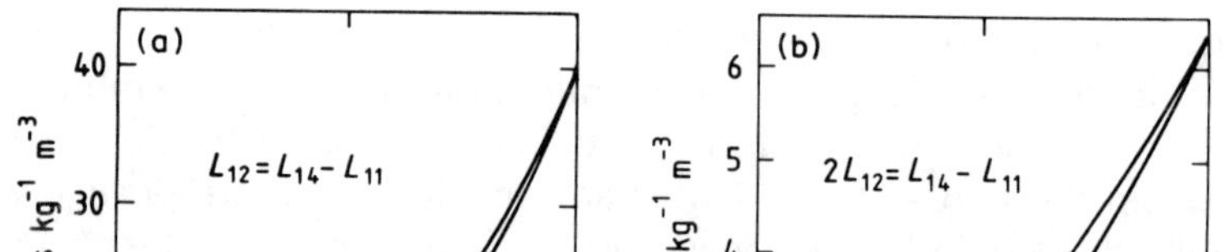

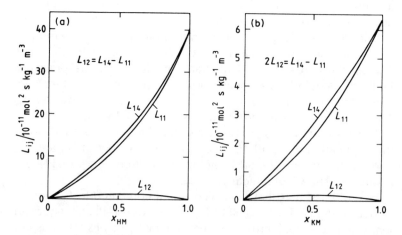

Figure 9.4. Variations in coefficients L_{12}, L_{11} and L_{14} with composition of membrane
(a) The system HM–NaM.
(b) The system KM–SrM_2.

9.3 *Transport of Water*

Water is carried through the membrane with ions, and it may also flow through the membrane separately. The flux equation for water is given by eqs. (5.54c) or (5.59c) when there are two salts in the aqueous solutions adjacent to an ion exchange membrane.

9.3.1 *Transference coefficient for water*

The transference coefficient for water, t_{H_2O} (t_3), is usually determined by measurements of streaming potential.[14] The streaming potential is defined by eq. (5.23) (see also eq. (5.72)).

Trivijitkasem and Østvold[5] report variations of t_{H_2O} with the charge of the cation in the membrane CR and with the concentration of the salt in the aqueous solutions in equilibrium with the membrane. Their results are shown in Fig. 9.5. It can be seen that in dilute solutions, $c < 0.1$ kmol m^{-3}, there is only a small variation in t_{H_2O} when the solution contains the salt of a monovalent

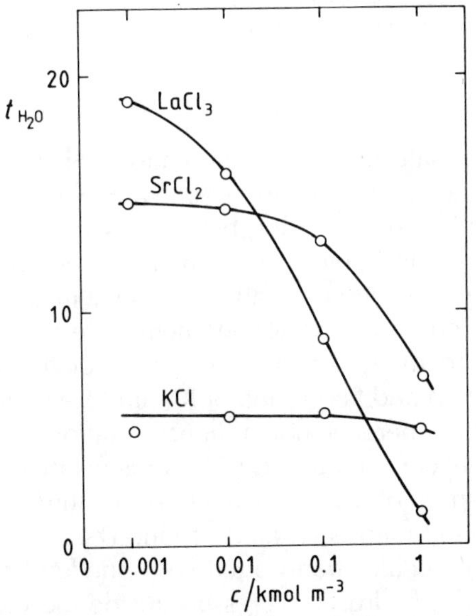

Figure 9.5. The transference coefficient for water in membrane CR as a function of salt concentration in adjacent aqueous solutions. Reprinted with permission from *Electrochimica Acta,* **25**, Trivijitkasem, P. and Østvold, T., Water transport in ion exchange membranes. Copyright (1980) Pergamon Journals Ltd.

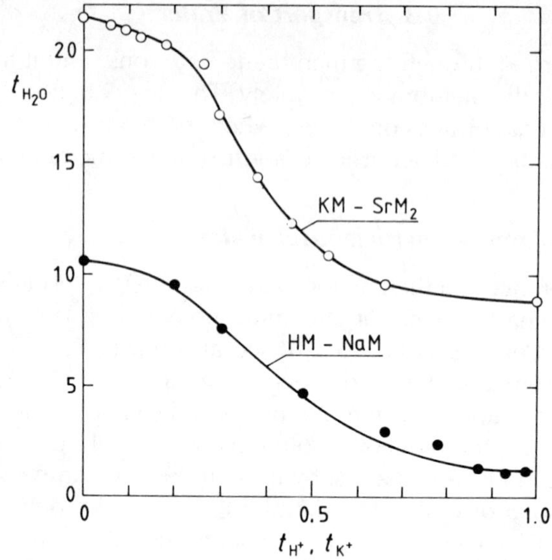

Figure 9.6. The transference coefficient for water, t_{H_2O}, as a function of the ionic transference number, t_{H^+} for the system HM – NaM and t_{K^+} for the system KM – SrM_2.

cation, KCl, or a salt of a divalent cation, $SrCl_2$. At higher concentration of salt in the solution, or with a trivalent cation present, the membrane may not behave as a perfect cation exchange membrane and anions may enter the membrane.

When there are two different cations present in a membrane, t_{H_2O} varies with composition. Experimental results from the measurement of streaming potential are given for the membrane CR as mixtures of HM and NaM[6] and of KM and SrM_2[3] in Fig. 9.6. It can be seen that there is not a linear relationship between t_{H_2O} and t_{H^+} or between t_{H_2O} and t_{K^+}. This means that we *cannot* interpret the transport of water as a constant number of water molecules transported with each kind of ion. The variation in the number of water molecules transported with one kind of ion may lead to a variation in hydrostatic pressure inside the membrane,[4] and thereby to local variations in the chemical potentials (including that of water). We have no report on the influence of this variation on the Nernst–Planck flux equations. The figure shows that t_{H_2O} is much higher for the pure SrM_2 membrane than for the pure

NaM membrane and the pure KM membrane. The lowest value of t_{H_2O} is found for the pure HM membrane. The water content of the membrane shows only a small variation with composition.[8]

9.3.2 *Osmotic and hydraulic water flux*

Osmotic phenomena were treated in Section 5.4.2. The water flux in the cation exchange membrane CR in contact with HCl solutions can be written as

$$J_{H_2O} = -\bar{l}_{H_2O}\,(\Delta\mu_{H_2O}(c) + V_{H_2O}\Delta p) \tag{9.15}$$

(Cf. eq. (5.28).) Anions of the aqueous solution do not enter the membrane, and for reasons of electroneutrality cations alone cannot migrate through the membrane. The water flux is the only flux through the membrane. The difference in chemical potential for water is determined by the difference in concentration for the solutions adjacent to the membrane.

The coefficient $\bar{l}_{H_2O}$ was determined by two independent experiments, the one by observing the flux of water as a function of $\Delta\mu_{H_2O}(c)$ when $\Delta p = 0$, the other by observing the flux of water as a function of Δp when $\Delta\mu_{H_2O}(c) = 0$.[6] The following results were obtained:

Osmotic flux of water, $\Delta p = 0$:

$$\bar{l}_{H_2O} = -J_{H_2O}/\Delta\mu_{H_2O}(c) = 3.0 \times 10^{-6}\ \text{mol}^2\ \text{s kg}^{-1}\ \text{m}^{-3} \tag{9.16}$$

Hydraulic flux of water, $\Delta\mu_{H_2O}(c) = 0$:

$$\bar{l}_{H_2O} = -J_{H_2O}/V_{H_2O}\Delta p = 5.9 \times 10^{-6}\ \text{mol}^2\ \text{s kg}^{-1}\ \text{m}^{-3} \tag{9.17}$$

According to eqs. (9.16) and (9.17), the pressure-dependent part of the chemical potential gives a larger flux of water than the concentration-dependent part. This is not expected from eq. (9.15). So far we have no explanation of the discrepancy.

Reverse osmosis, which is used for many practical processes, is based on the transport of water through the membrane under pressure, leaving the solutes behind.

9.4 *Conclusions*

Experimental data for transport of ions in the membrane show a consistency that gives credence to the values and to the theory.

The simple Nernst–Planck flux equations are useful in the quantitative description of fluxes, being good approximations to the more complex flux equations. In Section 8.1.1. it was seen that the Nernst–Planck equations represent good approximations for dilute aqueous solution, while they are inaccurate in more concentrated solutions.

Experimental data for transport of water are less conclusive than the data for transports of ions and more research is needed.

9.5 *References*

1. Lacey, R.E. and Loeb, S., *Industrial Processing with Membranes,* Krieger, Huntington, 1979.
2. Lior, N. (ed.), Measurement and control in water desalination, *Desalination 59* (1986), special issue, complete volume.
3. Holt, T., *Thesis,* Division of Physical Chemistry, University in Trondheim, N.T.H., 1983.
4. Holt, T., Ratkje, S.K., Førland, K.S. and Østvold, T., *J. Membr. Sci.,* **9**, 69 (1981).
5. Trivijitkasem, P. and Østvold, T., *Electrochim. Acta,* **25**, 171 (1980).
6. Skrede, M.G., *Thesis,* Division of Physical Chemistry, University in Trondheim, N.T.H., 1988.
7. Weber, W.J., *Physicochemical Processes for Water Quality Control,* Wiley–Interscience, New York, 1972.
8. Holt, T., Førland, T. and Ratkje, S.K., *J. Membr. Sci.,* **25**, 133 (1985).
9. Helfferich, F., *Ion Exchange,* McGraw-Hill, New York, 1962.
10. Harned, H.S. and Owen, B.B., *The Physical Chemistry of Electrolyte Solutions,* 3rd ed., Reinhold, New York, 1958.
11. Yeo, R.S. and Yeager, H.L., Structural and transport properties of perfluorinated ion-exchange membranes, in *Modern Aspects of Electrochemistry* (eds. Conway, B.E., White, R.E. and Bockris, J.O'M), Vol 16, Plenum Press, New York, 1985.
12. Paterson, R., Cameron, R.G. and Burke, I.S., Interpretation of membrane phenomena using irreversible thermodynamics, in *Charged Gels and Membranes* (ed. Sélégny, E.), Reidel, New York, 1976.
13. Kontturi, K., Ekman, A. and Forssell, P., *Acta Chem. Scand.,* **A39**, 273 (1985).
14. Brun, T.S. and Vaula, D., *Ber. Bunsenges. Physik. Chem.,* **71**, 824 (1967).

CHAPTER 10

Energy Conversion in Biological Systems

The nutrients taken in by humans and animals release energy by reaction with oxygen and the reaction products are mainly carbon dioxide and water. The overall reaction is a redox reaction involving transfer of electrons. It consists of a chain of successive steps catalysed by enzymes. Carbon dioxide is released and finally oxygen takes up electrons and water is formed.

We shall consider only the last few steps of this chain, the *oxidative phosphorylation*, where molecular oxygen is reduced and *adenosine triphosphate* (ATP) is formed. The oxidative phosphorylation takes place in the *mitochondria* which are *organelles*, membrane-surrounded bodies inside the cell. The membrane of the mitochondrion plays an important part in the oxidative phosphorylation. The ATP stores energy for processes such as *muscular contraction*, where the energy is converted to mechanical work, and other energy-requiring processes. The two processes, oxidative phosphorylation and muscular contraction, or similar processes, are central to all life.

We shall start with the simpler process, muscular contraction, in Section 10.1 and continue with the more complex process, oxidative phosphorylation, in Section 10.2.

The mechanism of muscular contraction is very much related to the mechanisms in other biological units where movement is created. On the molecular level there are similarities between the contractile process of a skeletal muscle and the propelling process of a sperm tail, a *flagellum*.[1] The motions of *cilia*, hair-like spikes extending in large numbers from some cells, are also similar. The cilia move extracellular material along the cell surface. Motions inside a cell are also created by units similar to those of a skeletal muscle.[2]

Mitochrondria, the 'power plants' of the cells, are present in almost all cells of higher organisms. In the cells of the leaves of green plants there is another organelle, the *chloroplast*, having a membrane with a similar function as the membrane of the mitochondrion. Here the energy of *light* is transformed into chemical energy in the form of ATP. In *bacteria* also one finds membranes with a similar function. Nicholls[3] treats the processes in the chloroplasts and in the purple bacteria.

10.1 *Muscular Contraction*

The muscle converts chemical energy into mechanical work. Efficiencies (mechanical work/energy input) of almost 0.45 have been reported,[4] hence the muscle competes favourably with the best steam engines. To achieve such a high efficiency, the energy conversion must operate close to equilibrium and the linear equations of irreversible thermodynamics may be applied to muscular contraction (cf. Sections 1.1. and 1.2).

The skeletal muscle is the type of muscle most extensively investigated. Based on the structure of the muscle, we shall seek a mechanism that can account for the high efficiency when the scalar energy of ATP creates a mechanical vectorial force. We shall avoid explanations involving high energy losses for single steps, as these would lead to a lower efficiency. Based on the model suggested for the mechanism and energetic data, we shall use the tools of irreversible thermodynamics to interpret the different types of energy losses in the muscle.

10.1.1 *Structure of skeletal muscles. Mechanism of muscular contraction. Energetics*

Skeletal muscles are built up from fibres that are attached to bones via tendons. The fibres are very long parallel cells containing the *myofibrils*, which can contract and expand. The myofibrils inside the fibres are surrounded by a fluid containing solutes—among others *adenosine triphosphate* (ATP), *adenosine diphosphate* (ADP), *phosphocreatine* (PCr) and inorganic salts. Organelles such as *sarcoplasmic reticula* and *mitochondria* are also present in the fluid. We shall study the function of the different constituents and start with a description of the myofibrils.

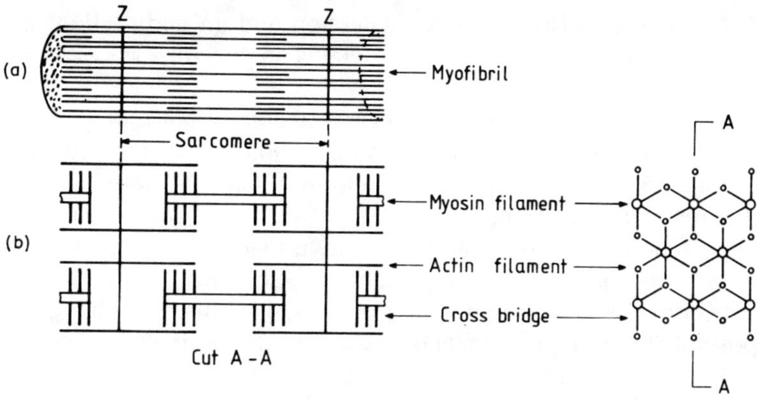

Figure 10.1. (a) The myofibril, cut. (b) The arrangement of myosin filaments and actin filaments in a myofibril.

A part of a myofibril is pictured in Fig. 10.1. The long myofibril cells are divided into sections, *sarcomeres*, by partitions, the Z lines.

The myofibrils contain two types of filaments, the *myosin* (thick) filaments and *actin* (thin) filaments. The two types of filaments overlap. When the muscle contracts, the sarcomeres are shortened by increasing overlap of the two types of filaments, and the Z–Z distances become shorter.[5,6] A more detailed picture of myosin and actin is given in Fig. 10.2.

A single myosin filament (Fig. 10.2(a)) is a bundle of long myosin molecules. A part of each molecule protrudes from the filament. This forms a bridge over to actin when the muscle performs work.

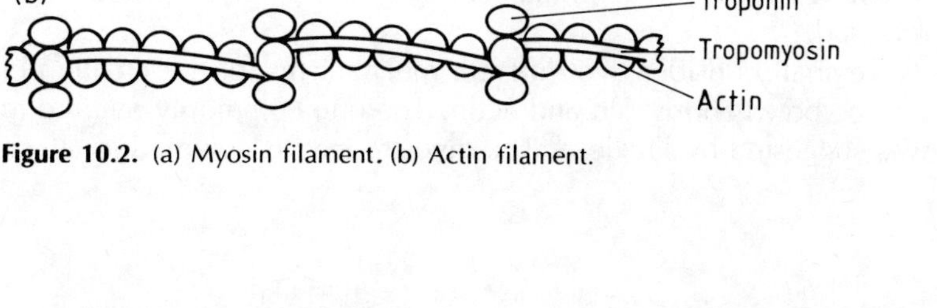

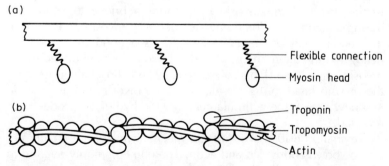

Figure 10.2. (a) Myosin filament. (b) Actin filament.

The bridge consists of a flexible section and an end section, the *myosin head*. The myosin head contains an enzyme catalysing the hydrolysis of ATP.

The actin filament (Fig. 10.2(b)) is a double helix of globular actin molecules. The filament also contains two other molecules at regular intervals, *tropomyosin* and *troponin*. They are active in the regulation of bond formation between a myosin head and actin. The tropomyosin blocks bridge formation when the muscle is in a resting state. In the presence of Ca^{2+} ions an interaction between Ca^{2+}, troponin and tropomyosin causes a change in position for tropomyosin allowing for bridge formation between myosin heads and actin.

The Ca^{2+} ions are released from the sarcoplasmic reticulum upon the triggering by a nerve signal. The mitochondria supply ATP, the source of energy for muscle contraction (see Sectin 10.2). Lehninger[7] gives more details on the structure of muscles.

When a muscle performs work, the myosin heads go through a cyclic process, where one ATP molecule is hydrolysed per myosin head for each cycle. The ATP is hydrolysed to give one ADP and an inorganic phosphate releasing energy:

$$
\begin{array}{c}
\text{O}^- \quad\ \text{O}^- \quad\ \text{O}^- \qquad\qquad\qquad \text{O}^- \quad\ \text{O}^- \\
|\qquad\ \ |\qquad\ \ |\qquad\qquad\qquad\quad |\qquad\ \ | \\
\text{R}-\text{O}-\text{P}-\text{O}-\text{P}-\text{O}-\text{P}-\text{O}^- + \text{H}_2\text{O} \rightarrow \text{R}-\text{O}-\text{P}-\text{O}-\text{P}-\text{O}^- + \text{H}_2\text{PO}_4{}^- \\
\|\qquad\ \ \|\qquad\ \ \|\qquad\qquad\qquad\quad \|\qquad\ \ \| \\
\text{O}\qquad\ \text{O}\qquad\ \text{O}\qquad\qquad\qquad\quad \text{O}\qquad\ \text{O}
\end{array}
$$

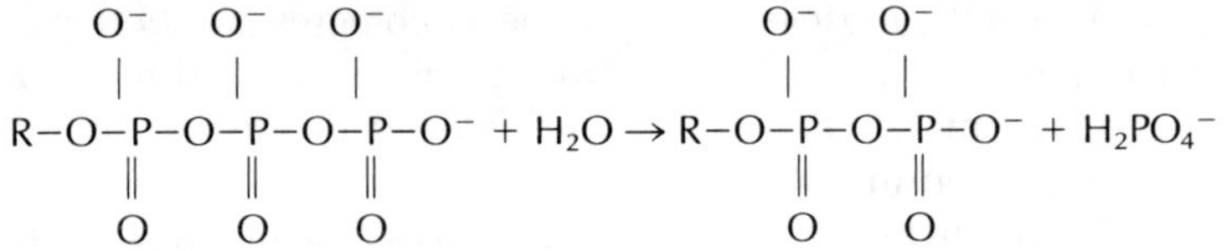

$$(10.1)$$

Adenosine triphosphate   Adenosine diphosphate

The cyclic process can be considered as consisting of two parts. In the first part the myosin head forms a bridge to actin, and during attachment a force operates between myosin and actin. In the second part of the cycle ATP forms a bond to the myosin head, and the actin–myosin bond is broken. While bonded to the myosin head, the ATP hydrolyses to ADP and phosphate. Thereafter the myosin head forms a bridge between myosin and actin again at a new site on the actin filament and the hydrolysis products are released.[8]

We shall consider two different mechanisms for the creation of a force between myosin and actin. The one commonly referred to was suggested by Huxley.[9] The myosin head is assumed to attach

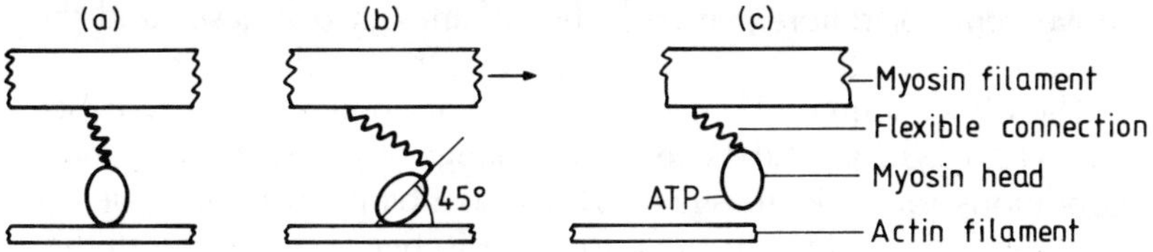

Figure 10.3. Force generation by rotation of myosin head; (a) bond formation, (b) 45° rotation, (c) bond broken by ATP.

to actin at a right angle, and then, while still attached, it makes a 45° rotation, thereby creating a force (see Fig. 10.3). More recent studies of spin labels attached to cross bridges,[10] however, indicate that the myosin head does not change orientation while exerting force.

An alternative explanation of force creation was suggested by Førland.[11] After attachment the myosin head is assumed to move along an interval between two troponin sites on the actin filament without changing the angle between the myosin head and the actin filament (see Fig. 10.4). The figure pictures a movement of a myosin head from left to right along the actin filament. A myosin head attaches itself to actin at a position A near the lefthand side of the interval. By stretching and bending the elastic part of the cross bridge the myosin head moves towards the right forming stepwise stronger bonds to actin. This exerts a force, which moves the myosin filament relative to the actin filament. At position B the myosin head has arrived at the point of maximum bond strength. Here the bond between myosin and actin must be broken in order that the myosin proceed to the next interval. The

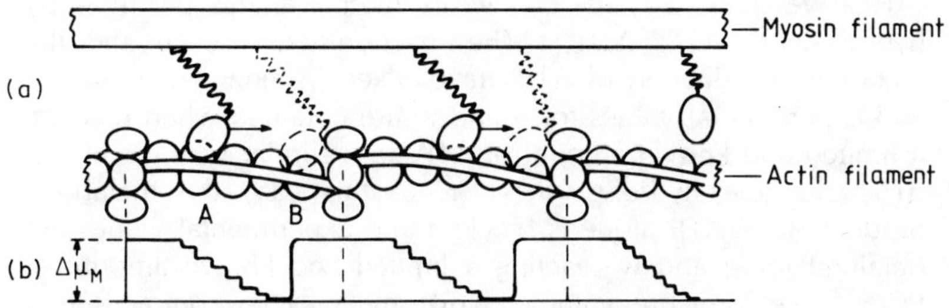

Figure 10.4. (a) Force generation by moving myosin heads along an actin filament. (b) Chemical potential difference of myosin heads during movement.

movements of different myosin heads are out of phase, and the result is a smooth release of energy in very small steps.

The direct source of energy in muscular contraction is ATP (see eq. (10.1)). The Gibbs energy change by hydrolysis under conditions found in muscles, ΔG_{ATP}, is obtained from studies of equilibria in a complete cycle for myosin heads. From the studies by Taylor[12] and Lowe[8] values of -57 kJ mol^{-1} and -67 kJ mol^{-1}, respectively, are obtained.

When a muscle performs work and ATP is consumed, the concentration of ATP is still close to constant.[13] The consumed ATP is rapidly restored by a reaction between ADP and phosphocreatine (PCr):

$$
\begin{array}{cc}
\text{O}^- \qquad \text{CH}_3 & \text{H} \quad \text{CH}_3 \\
| \qquad\quad | & | \quad\; | \\
\text{ADP} + {}^-\text{O}-\text{P}-\text{N}-\text{C}-\text{N}-\text{CH}_2-\text{COO}^- \rightarrow \text{ATP} + {}^+\text{H}-\text{N}-\text{C}-\text{N}-\text{CH}_2-\text{COO}^- \\
\;\;\| \; | \; \| & \qquad | \; \| \\
\;\;\text{O} \; \text{H NH} & \quad \text{H NH}
\end{array}
$$

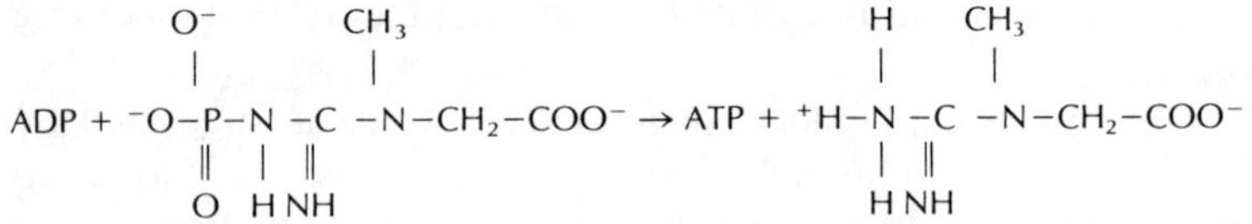

$$(10.2)$$

Phosphocreatine   Creatine

The concentration of PCr in a muscle is much higher than the concentration of ATP, and PCr represents a way of storing energy ready for use. In energy calculations it is common to consider PCr as the source of energy. The Gibbs energy change for the hydrolysis of PCr, the sum of the reactions (eqs.(10.1) and (10.2)) under the conditions in the muscle, cannot easily be obtained directly. Changes in internal energy, ΔU, however, can be obtained from observed values of heat, q, and work, w, when a muscle contracts. Section 10.1.3 shows how this can be utilized to obtain a value for the chemical potential of the myosin heads.

First we shall study the change in internal energy per mole by hydrolysis of ATP, ΔU_{ATP}. The reaction (eq. (10.2)) and the metabolic production of ATP are blocked by enzyme poisoning and experimental values for q and w are obtained when reacting a limited and known amount of ATP. In a similar way we obtain ΔU_{PCr} for the hydrolysis of PCr. In this case the metabolic production of ATP alone is blocked and experimental values are obtained for q and w reacting a limited and known amount of PCr. Since pressure–volume work by muscle contraction is negligible, changes in internal energy can be replaced by enthalpy changes, $\Delta U_{ATP} \approx \Delta H_{ATP}$ and $\Delta U_{PCr} \approx \Delta H_{PCr}$.

The more accurate determinations of q and w were obtained with PCr as the source of energy,[13] and the result is

$$\Delta U_{PCr} = \Delta H_{PCr} = -46.4 \text{ kJ mol}^{-1} \tag{10.3}$$

Within the limits of error ΔU_{PCr} and ΔU_{ATP} are equal.[14]

10.1.2 *Application of irreversible thermodynamics*

When a muscle performs work, metabolic energy is expended in the transport of myosin and actin filaments relative to each other. Irreversible thermodynamics can be used to describe the transports.

In the stationary state transport equations are simplified since forces can be expressed by differences instead of gradients, which in many cases are not known in sufficient detail. Although the contraction of the myofibril cannot be a true stationary process, it can with a good approximation be treated as such. Caplan and Essig[15] summarize the evidence that the process can be considered as *quasi-stationary*, i.e. reservoirs are large compared to changes during transport.

Caplan and Essig apply the irreversible thermodynamics without referring to any particular mechanism. They consider the output of mechanical work and the progress of the chemical reaction coupled to the mechanical process of muscle contraction. They assume that it is possible to identify one unique chemical reaction coupled to the mechanical process and that this reaction is not coupled directly to any of the other chemical reactions taking place. The dissipated energy per unit time for this process only can be written as follows:

$$\frac{TdS}{dt} = v(-Ten) + J_r A \tag{10.4}$$

where v is the contraction velocity for the muscle, Ten is the tension of the muscle, J_r is the velocity of the chemical reaction and A (= $-\Delta G$) is the affinity of the chemical reaction coupled to the mechanical process. The corresponding flux equations are

$$v = L_c(-Ten) + L_{cr}A \tag{10.5a}$$

$$J_r = L_{cr}(-Ten) + L_r A \tag{10.5b}$$

where L_c, L_{cr} and L_r are phenomenological coefficients and the subscripts c and r indicate contraction and reaction, respectively.

The flux equations contain both scalar and vectorial quantities. According to the Curie principle (Section 3.4) there is no coupling between scalar and vectorial quantities in an isotropic system. Since a vectorial contraction velocity is obtained from the scalar force, affinity, it means that the coupling is caused by an anisotropy in the system and the coefficient L_{cr} is a vector. The equations describe the process on a macroscopic level. A molecular interpretation of anisotropy and the vectorial coefficient is not dealt with.

Experimental results for muscle contraction indicate that the value of A is variable and that the coupling between the mechanical process and the chemical process is incomplete. It is difficult to explain a significant variation in A for *one* chemical reaction unless relations between activities change. Caplan and Essig[15] discuss the nature of a regulatory process responsible for the incomplete coupling and point out the continuous breakdown of ATP during *isometric tetanus*, where the muscle is under tension at constant length, $v = 0$, and no work is performed.

We shall use a different approach when applying irreversible thermodynamics. The derivation will be based on the mechanism involving moving myosin heads (discussed in Section 10.1.1). A myosin head attached to the actin filament moves in a stepwise manner along an interval on the actin filament because the strength of the bond to actin increases for each step. The change in bond strength over the interval can be expressed as a change in chemical potential of the myosin head, $\Delta\mu_M$. In this first part of the cyclic process for myosin heads, work is performed.

Some of the energy in the first part of the cycle is dissipated as heat. In the second part of the cycle, where the actin–myosin bridge is broken, all the energy is dissipated as heat. As was mentioned, ATP is continually consumed during isometric tetanus, dissipating the energy as heat. We may assume that this process continues in *parallel* with the work-producing process, involving some of the myosin heads. The process is discussed in more detail in Section 10.1.3.

We shall apply irreversible thermodynamics separately to the first part of the cyclic process for the myosin heads, the part where work is performed. Our system, pictured schematically in Fig. 10.4, consists of the right-hand side halves of the sarcomere pictured in

Fig. 10.1. We have chosen the actin filament as the frame of reference for transport. The myosin heads (and the myosin filament) move to the right when work is performed. The change in chemical potential per mole myosin heads, $\Delta\mu_M$, over an interval of length, ℓ, gives an average force $\overline{X}_M = -\Delta\mu_M/\ell$. The conjugate flux, J_M, refers to the movement of one mole myosin heads attached to the actin filament. This flux is equal to the velocity of the myosin filament relative to the neighbouring actin filament, and it is also equal to the contraction velocity for the muscle, $J_M = v$. When a muscle carries out work $(-w)$, a mechanical force pulls the myosin heads in the opposite direction of $\overline{X}_M$. The average mechanical force over the interval ℓ is w/ℓ. This force operates on one mole myosin heads, and the conjugate flux is $J_M = v$.

In the stationary state it may be convenient to use potential differences as forces instead of average forces over the length of a system (see Section 3.5). Using $-\Delta\mu_M$ and w as forces, both given as joule per mole attached myosin heads, we obtain the dissipated energy per mole attached myosin heads per unit time:

$$\frac{TdS}{dt} = v\,w + J_M\,(-\Delta\mu_M) \tag{10.6}$$

or, since $J_M = v$:

$$\frac{TdS}{dt} = J_M\,(w - \Delta\mu_M) \tag{10.7}$$

with the corresponding flux equation:

$$J_M = v = L\,(w - \Delta\mu_M) \tag{10.8}$$

The two expressions for dissipated energy per unit time, eqs. (10.4) and (10.6), may be compared. The first term on the right-hand side is the same in both equations. The second term, however, is different in the two equations. The term J_rA refers to the chemical reaction furnishing the energy for the process, while $J_M\,(-\Delta\mu_M)$ refers to the moving myosin heads only. When there is a parallel energy-consuming process involving myosin heads that do not move, J_rA represents two independent processes which are not linearly related.

The flux equations (eqs. (10.5)) combine a vectorial flux, $\boldsymbol{v}$, with a scalar force, A, and a scalar flux, J_r, with a vectorial force, $(-\, Ten)$, by means of a vectorial cross coefficient, L_{cr}. Both flux and force in eq. (10.8) are vectors and the coefficient, L, is a scalar.

Introducing a *resistance coefficient*, $R = 1/L$, into eq. (10.8) we obtain

$$R\boldsymbol{v} = \mathrm{w} - \Delta\mu_{\mathrm{M}} \tag{10.9}$$

The linear relation in eq. (10.9) is expected to be valid when $(w - \Delta\mu_{\mathrm{M}})/N_{\mathrm{A}}$ per step is small compared to kT (see Section 1.2). Here N_{A} is the Avogadro constant, the number of myosin heads in one mole.

10.1.3 *Energy balance for a complete cycle for myosin heads*

The internal energy released by the hydrolysis of one mole PCr (eq. (10.3)) is converted to heat and work by a complete cycle for one mole myosin heads, forming a bridge to actin, moving along the actin filament and being released from actin again. We shall consider the consequences of eq. (10.8) when analysing the energy balance and use the results obtained by Bendall[14] on heat and work as functions of contraction velocity for a *frog sartorius muscle*.

The energy change for the complete cycle is the sum of the energy changes for the first part of the cycle, ΔU_{I}, and for the second part, ΔU_{II}.

$$\Delta U_{\mathrm{PCr}} = \Delta U_{\mathrm{I}} + \Delta U_{\mathrm{II}} \tag{10.10}$$

For the first part of the cycle $\Delta\mu_{\mathrm{M}} = w - R\boldsymbol{v} = \Delta G_{\mathrm{I}} \approx \Delta U_1 - T\Delta S_{\mathrm{M}}$. The dissipated energy per mole moving myosin heads is $R\boldsymbol{v}$ and the corresponding absorbed heat is q_1, a linear function of contraction velocity:

$$q_1 = -R\boldsymbol{v} \tag{10.11}$$

The term $T\Delta S_{\mathrm{M}}$ involves the change in partial molar entropy for myosin heads when moving from position A on the actin to position B (see Fig. 10.4). The heat absorbed is $q' = T\Delta S_{\mathrm{M}}$. This heat is independent of velocity.

In the second part of the cycle no work is performed, and the whole energy ΔU_{II} is converted to heat, which is also independent of velocity.

The total heat absorbed by the cyclic process is

$$q_1 + q' + \Delta U_{II} = q_1 + q_2 = -Rv + q_2 \tag{10.12}$$

The velocity-independent parts of the total heat absorbed, $q' + \Delta U_{II}$, are collected in the one term, q_2:

$$q_2 = \text{const} \tag{10.13}$$

For the highest velocity, where $w = 0$, we have

$$\Delta U_{PCr} = -Rv_{max} + q_2 \tag{10.14}$$

We shall consider another cyclic process in parallel with the work-performing process. The parallel process consumes energy, but does not contribute to the work performed by the muscle. As mentioned, a muscle consumes ATP (and thereby PCr) during isometric tetanus, when no work is performed. This may be interpreted as a breaking and re-forming of an actin–myosin bridge at the same position B on the actin filament consuming one ATP per cycle. We may assume that this process will also occur for low values of v, involving gradually fewer myosin heads as v increases. For higher values of v, a myosin head released from actin at position B will move some distance to the right of B before forming a bridge again, and bridge formation at the same position B is less probable. The dissipated energy due to reattachment at B is expected to decrease rapidly with increasing v.

The mole fractions of myosin heads going through the first and second process are x_1 and x_2 respectively. The total heat absorbed per mole PCr consumed is

$$q = x_1(-Rv + q_2) + x_2\Delta U_{PCr}$$

or

$$q = x_1(-Rv + q_2) + x_2(-Rv_{max}) + x_2 q_2 \tag{10.15}$$

The term $x_2(-Rv_{max}) = q_3$ is the additional heat absorption caused by the second process. It represents a loss in energy since it cannot be converted to work. We have

$$q_3 = x_2(-Rv_{max}) = q - q_2 + x_1 Rv \tag{10.16}$$

With increasing contraction velocity, x_2 decreases and thus $q_3 \rightarrow 0$. In the limit eq. (10.15) becomes identical to eq. (10.12).

The dissipation of energy caused by the velocity of the fluid surrounding the filaments may be neglected.[11]

The energy balance for the whole process is

$$\Delta U_{PCr} = q_1 + q_2 + q_3 + w = -46.4 \text{ kJ mol}^{-1} \tag{10.17}$$

The three contributions to the dissipation of energy are separable because of their different dependence on v. Figure 10.5 shows how they add up to the total heat absorption for varying contraction velocity in the frog sartorius muscle. At zero velocity q_1 vanishes and at high velocity q_3 vanishes. In the ideal case the curve for the total heat produced should approach a straight line at higher velocity, with the slope equal to R, the same slope as the line for q_1. The slight deviation in the curve at the highest contraction velocities represents a loss in energy. At very high velocities some myosin heads may not be released quickly enough from the actin in position B. The rate of release is of the order of magnitude 10^3 s^{-1}.[16] A myosin head that remains in position B, while the myosin filament moves on to the right, will impede the movement of the filament and cause a loss in energy.

From Fig. 10.5 the following values can be read. The slope of the slanting line gives $q_1 = -1.3\ v$ kJ mol^{-1} where the dimension for v is nanometre per millisecond, i.e. 10^{-6} m s^{-1}. The intercept with the ordinate for $v = 0$ gives $q_2 = -14$ kJ mol^{-1}. The heat q_3 is the difference between the total heat, q, and $q_1 + q_2$. Myosin heads contributing to q_3 do not contribute to q_1. This leads to a small correction in both q_1 and q_3,[11] not included here. The heat evolved, $-q_3$, decreases approximately exponentially with increasing v. The $\Delta\mu_M$ is a state function and thus of constant value. For the highest velocity where $w = 0$, we obtain from eq. (10.9) $\Delta\mu_M = -Rv_{max} = -32.4$ kJ mol^{-1}.

The crucial point in the above analysis is the slope of the line for q_1 in Fig. 10.5, as there is some uncertainty in drawing the line. We shall therefore seek information about $\Delta\mu_M$ from other experiments independent of the one discussed above.

Taylor[12] investigated equilibria in solution for all steps in the cyclic process of myosin heads. He suggests that the performance of work is associated with a particular series of transitions.

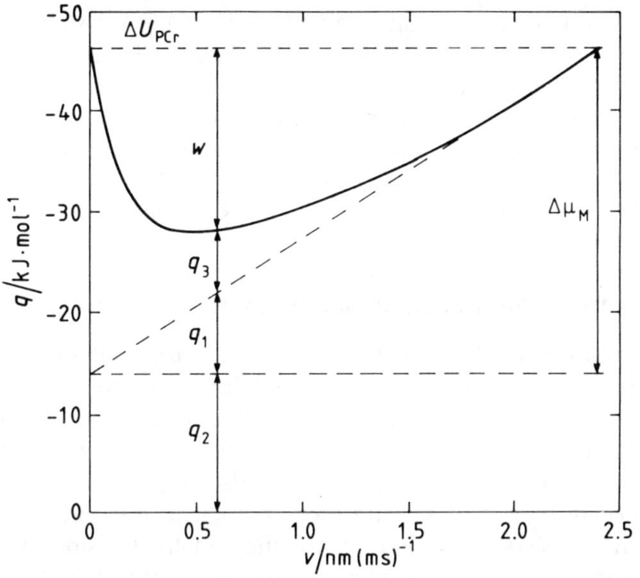

Figure 10.5. Absorbed heat as a function of contraction velocity — experimental curve[11, 14] (see text for explanations of symbols and numerical values). Reproduced from the *Biophysical Journal* (1985), **47**, pp. 665–671, by copyright permission of the Biophysical Society.

From his experimental equilibrium constants we can obtain $\Delta G = -34$ kJ mol^{-1} for the series. This is close to the above value, $\Delta\mu_M = -32.4$ kJ mol^{-1}, for the first part of the cyclic process.

There is a correlation between $\Delta\mu_M$ and the maximum *isometric tension* for the muscle, i.e. the maximum load per unit area that the muscle can hold without increasing its length. Calculations from maximum isometric tension for a frog sartorius muscle give a value $\Delta\mu_M = -36$ kJ mol^{-1},[11] which again is not very far from the value obtained from Fig. 10.5.

10.1.4 *Conclusions*

A specific mechanism for muscle contraction was used as the basis for formulating equations of irreversible thermodynamics. In principle the irreversible thermodynamics is independent of mechanism. It cannot prove any specific mechanism, but it can exclude some mechanisms as impossible.

Irreversible thermodynamics may reveal relations between properties. In the present case we found the value for $\Delta\mu_M$ from the maximum velocity and we were able to confirm this value by means of results from independent experiments.

Irreversible thermodynamics helps to explore the consequences of a proposed mechanism and it can be a useful tool for developing and checking theories about mechanisms in biological processes.

10.2 *The Mitochondrion as an Energy Converter*

Mitochondria are globular structures about 1 μm in diameter. They are found in the *cytosol*, the viscous solution inside the cell wall. They contain an enzyme-rich solution, the *matrix*, inside a double membrane, the *inner membrane* and the *outer membrane*. The inner membrane consists of phospholipids and proteins (see Section 5.1. concerning the structure of biological membranes). It is extensively folded compared to the relatively smooth outer membrane. A mitochondrion is pictured schematically in Fig. 10.6.

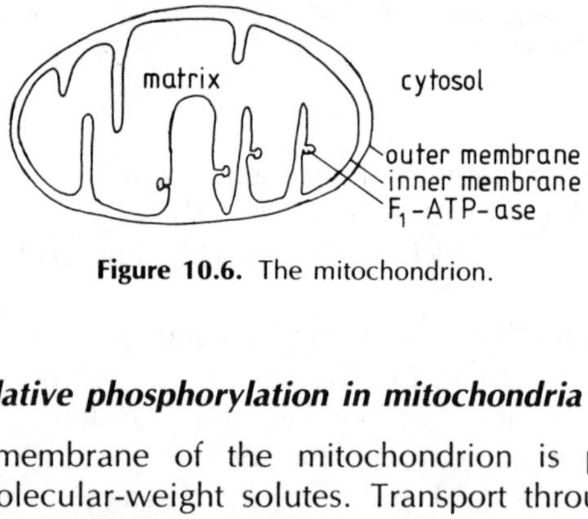

Figure 10.6. The mitochondrion.

10.2.1 *Oxidative phosphorylation in mitochondria*

The outer membrane of the mitochondrion is permeable to most low-molecular-weight solutes. Transport through the inner membrane, however, is restricted, as we shall see later. *Metabolites*, chemical intermediates from nutrients, are transported into the matrix where they react with *nicotinamide adenine dinucleotide* (NAD^+), present in the matrix, to form CO_2 and NADH as the major products.

The final steps in the oxidation of nutrients take place inside the inner membrane. Here NADH loses two electrons and forms NAD^+ and H^+:

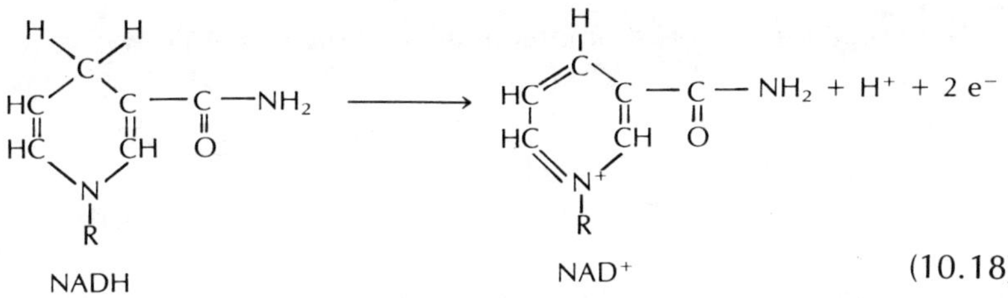

(10.18)

The electrons go through a chain of three steps, called the *respiratory chain*, and are finally received by oxygen. The reaction can be written as follows:

$$\mathrm{NADH(in)} + \mathrm{H^+(in)} + \tfrac{1}{2}\,\mathrm{O_2(in)} \rightarrow \mathrm{NAD^+(in)} + \mathrm{H_2O(in)} \tag{10.19}$$

where (in) means inside the membrane.

For each step in the respiratory chain approximately 4 H^+ are transported from the inside to the outside of the membrane.[17] The mechanism of this process is not known. For each step we can write formally as follows:

$$4\ \mathrm{H^+\ (in)} \rightarrow 4\ \mathrm{H^+\ (out)} \tag{10.20}$$

where (out) means outside the inner membrane. The respiratory chain is illustrated in Fig. 10.7.

Simultaneously approximately three molecules of *adenosine triphosphate* (ATP) are formed for each two electrons going through the respiratory chain.[17]

One may assume the stationary state for the reaction. In the stationary state there can be no accumulation of charge in any part of the system and the positive charge transferred across the membrane (eq. (10.20)) must of necessity be brought back to the matrix. This takes place in connection with the synthesis of ATP. Three simultaneous transport processes are associated with the synthesis of ATP, given by eqs. (10.21)–(10.23). Two of these processes involve a net transport of positive charge from the outside of the inner membrane to the inside.

The folds of the inner membrane contain some knob-like protuberances consisting of the enzyme F_1-ATP-ase. Here the ATP is synthesized (see Fig. 10.7).

The equation for the synthesis may be written as follows:

$$3\ H^+(\text{out}) + ADP^{3-}(\text{in}) + HPO_4^{\ 2-}(\text{in}) \rightarrow ATPH_2^{\ 2-} + H_2O$$
$$\rightarrow ATP^{4-}(\text{in}) + 2\ H^+(\text{in}) + H_2O(\text{in}) \qquad (10.21)$$

The three H^+ migrate from the outside through a proton-conducting channel up to the interface with the F_1-ATP-ase, marked I–I in Fig. 10.7. The F_1-ATP-ase is impermeable to protons, but permeable to ADP^{3-}, ATP^{4-} and $HPO_4^{\ 2-}$. There must be a change of charge carrier at the interface, and the reaction takes place here.[18] One H^+ is used to form the new oxygen bridge giving ATP, $P{-}O^- + H^+ + HO{-}P \rightarrow P{-}O{-}P + H_2O$. Two more H^+ are linked to ATP when passing through the F_1-ATP-ase, giving the intermediate $ATPH_2^{\ 2-}$. Details of the process are unknown.

By the reaction (eq. (10.21)), H^+ is transported through the membrane, while the two other transport processes concern exchange of ADP and ATP across the membrane, and the transport of $HPO_4^{\ 2-}$ across the membrane.

The transfer of the fourth positive charge to the matrix is associated with the exchange of ADP and ATP, as pictured in Fig. 10.7:

$$ADP^{3-}(\text{out}) + ATP^{4-}(\text{in}) \rightarrow ADP^{3-}(\text{in}) + ATP^{4-}(\text{out}) \qquad (10.22)$$

The two ions are transported by a common carrier in the inner membrane. The ATP produced according to eq. (10.21) is brought out of the mitochondrion, while the reactant ADP is brought into the mitochondrion.

Inorganic phosphate is needed inside the mitochondrion for the synthesis of ATP by eq. (10.21). It is assumed to be exchanged with OH^- by means of a common carrier, as pictured in Fig. 10.7:

$$H_2PO_4^{\ -}(\text{out}) + OH^-(\text{in}) \rightarrow H_2PO_4^{\ -}(\text{in}) + OH^-(\text{out}) \qquad (10.23)$$

There is no net charge transfer by this process; the driving force is the difference in chemical potential.

The solutions both inside and outside the mitochondrion are buffered by the acid–base pair $H_2PO_4^{\ -}$–$HPO_4^{\ 2-}$ and production and consumption of H^+ or OH^- give only small changes in pH.

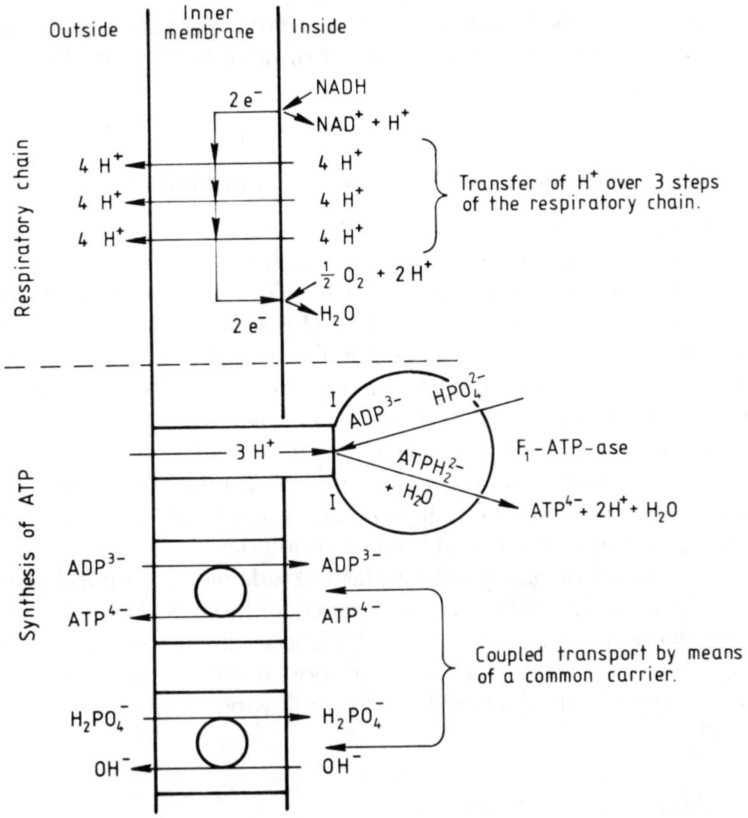

Figure 10.7. Oxidative phosphorylation.

Although ATP is produced inside the mitochondrion, the result of all reactions and transfers associated with the phosphorylation of ADP is as follows:

$$ADP^{3-}(\text{out}) + H_2PO_4^{-}(\text{out}) \rightarrow ATP^{4-}(\text{out}) + H_2O \qquad (10.24)$$

The combination of the respiratory chain and the phosphorylation of ADP is called the *oxidative phosphorylation*.

The main principle in the above description of oxidative phosphorylation, the transfer of protons from inside to outside by the oxidation of NADH and the synthesis of ATP by transferring protons back, is called the *chemiosmotic hypothesis*. It was suggested by Mitchell.[19] The number of protons transferred is still under dispute.

10.2.2 *Gibbs energies and the number of protons*

Lehninger[17] reports the Gibbs energy changes for the two major reactions in the oxidative phosphorylation. The Gibbs energy change for the reaction in eq. (10.19), under the conditions inside the mitochondrion, is $\Delta G_{ox} = -210$ kJ per mole NADH oxidized, and the Gibbs energy change for the reaction in eq. (10.24), under the conditions in the cell outside the mitochondrion, is $\Delta G_{ph} = 67$ kJ per mole ADP phosphorylated (cf. Lowe[8]).

For each mole NADH oxidized three moles of ATP are produced. This corresponds to an efficiency in energy conversion of $3 \times 67/210 = 0.96$. The high efficiency means that the conversion takes place in close to equilibrium conditions.

The transfer of protons is central in oxidative phosphorylation. The electrochemical potential difference for protons, $\Delta\tilde{\mu}_{H^+}$, is used extensively in thermodynamic calculations. In the application of irreversible thermodynamics to transport processes $\Delta\tilde{\mu}_{H^+}$ is used as a force.[3,19] It is defined by the following equations (cf. eq. (4.33)):

$$\Delta\tilde{\mu}_{H^+} = \Delta\mu_{H^+} + \Delta\psi \tag{10.25}$$

where $\Delta\mu_{H^+}$ is the chemical potential difference for H^+ and $\Delta\psi$ is the electrostatic potential difference over the membrane (the membrane potential). It was discussed in Section 4.1.2 that both quantities are in principle unmeasurable. In an ideal solution where activity coefficients are equal to unity, however, we may take $\mu_{H^+} - \mu^{o}_{H^+} = RT\ln c_{H^+}$, and similar for other ions.

We can find $\Delta\tilde{\mu}_{H^+}$ from two different measurements.[3,20] The difference in pH over the membrane gives

$$\mu_{H^+}(\text{in}) - \mu_{H^+}(\text{out}) = \Delta\mu_{H^+} = -\,2.3\frac{RT}{F}\Delta\text{pH} \tag{10.26}$$

The value of $\Delta\psi$ may be found from the equilibrium distribution of potassium ions across the membrane in presence of valinomycin as a carrier for K^+. When the potassium ions are in equilibrium we have

$$\tilde{\mu}_{K^+}(\text{in}) - \tilde{\mu}_{K^+}(\text{out}) = \Delta\tilde{\mu}_{K^+} = 0 = \Delta\psi + 2.3\frac{RT}{F}\log\frac{c_{K^+}(\text{in})}{c_{K^+}(\text{out})}$$

$$\Delta\psi = 2.3\frac{RT}{F}\log\{c_{K^+}(\text{in})/c_{K^+}(\text{out})\} \qquad (10.27)$$

In Section 4.1.2 the gradient $\nabla\tilde{\mu}_{H^+} = \nabla\varphi'$ was defined (eq. (4.39)). Correspondingly

$$\Delta\tilde{\mu}_{H^+} = \Delta\varphi' \qquad (10.28)$$

where $\Delta\varphi'$ is the well-defined emf of an electrochemical cell with two hydrogen electrodes. This operational definition of $\Delta\tilde{\mu}_{H^+}$ in mitochondria was applied by Førland and coworkers.[18]

Different values have been reported for $\Delta\tilde{\mu}_{H^+}$. Differences are due partly to uncertainties in measuring technique and partly to different states of the mitochondria. Nicholls[21] reported values in the range from −228 mV to −170 mV. The values depend on the concentration of ADP. For very low concentrations of ADP phosphorylation does not take place and $\Delta\tilde{\mu}_{H^+} = -228$ mV. The oxidation reaction, however, continues at a very low rate, because the membrane is not perfectly impermeable to H^+ which leaks back to the inside. There is some loss in energy when H^+ ions are transferred to the outside by the oxidation reaction, reducing the numerical value of $\Delta\tilde{\mu}_{H^+}$. When phosphorylation takes place, the transfer of H^+ by the oxidation reaction will increase, and the reduction in the numerical value will be larger.

The electrochemical potential, $\Delta\tilde{\mu}_{H^+}$, is related to the Gibbs energies, ΔG_{ox} and ΔG_{ph}, by stoichiometric coefficients. Under equilibrium conditions (no loss by transport) we have

$$\Delta G_{ox} = n\Delta\tilde{\mu}_{H^+} \qquad (10.29)$$

$$\Delta G_{ph} = -m\Delta\tilde{\mu}_{H^+} \qquad (10.30)$$

where n is the number of moles H^+ transported through the membrane from inside to outside for each mole NADH oxidized, and m is the number of moles H^+ transported from outside to inside for each mole ADP phosphorylated.

For $n = 12$ and $\Delta G_{ox} = -210$ kJ mol^{-1} we obtain

$$\Delta\tilde{\mu}_{H^+} = \frac{-210 \times 1000}{12 \times 96{,}500} = -\,0.18\ \text{V} \quad \text{or} -\,180\ \text{mV}$$

For $m = 4$ and $\Delta G_{ph} = 67$ kJ mol^{-1} we obtain

$$\Delta\tilde{\mu}_{H^+} = -\frac{67 \times 1000}{4 \times 96{,}500} = -\,0.17\ \text{V} \quad \text{or} -\,170\ \text{mV}$$

There is a fairly good agreement between these two values for $\Delta\tilde{\mu}_{H^+}$, but they do not agree with values of -228 mV as reported by Nicholls. In recent years evidences for a lower value of n have been published. Beavis and Lehninger[22,23] suggest that four H^+ are transported through the membrane in the first and in the second step of the respiratory chain, while only three H^+ are transported in the third step, giving $n = 11$. For $n = 11$, we obtain $\Delta\tilde{\mu}_{H^+} = -200$ mV. When $m = 4$, $11/4 = 2.75$ moles ADP are phosphorylated per mole NADH oxidized.

There are still uncertainties about ΔG_{ox}, ΔG_{ph}, n, m, and $\Delta\tilde{\mu}_{H^+}$. There are also substantial uncertainties about the mechanism.

10.2.3 *Application of irreversible thermodynamics*

The oxidative phosphorylation consists of several coupled processes, some in series and some in parallel. The mechanism is still unknown for several steps in the total process. The thermodynamic analysis indicates that the energy conversion takes place close to equilibrium and we may expect fluxes to be linear functions of forces.

The equations of irreversible thermodynamics can be written independent of mechanism. We may observe total forces (potential differences) and fluxes over unknown steps and examine whether the fluxes are linear functions of the forces. Scalar forces and fluxes may have to be used even though the process involves transports over potential gradients. A treatment based on vectorial forces and fluxes would require more detailed knowledge about mechanism.

Caplan and Essig[15] apply irreversible thermodynamics in the analysis of energy conversion in mitochondria. They use the forces, $\Delta\tilde{\mu}_{H^+}$, and the affinities, $A_{ph} = -\Delta G_{ph}$ and $A_{ox} = -\Delta G_{ox}$. The conjugate fluxes are J_{H^+}, J_{ph} and J_{ox}. In the stationary state, and with no flux of other ions, the net transport of protons is equal to zero. We are left with two fluxes and two forces. The dissipated energy per unit time can be expressed as follows:

$$T\frac{dS}{dt} = J_{ph}A_{ph} + J_{ox}A_{ox} \tag{10.31}$$

The corresponding flux equations are

$$J_{ph} = L_{11}A_{ph} + L_{12}A_{ox} \tag{10.32a}$$

$$J_{ox} = L_{21}A_{ph} + L_{22}A_{ox} \tag{10.32b}$$

The relations in eqs. (10.32) were investigated by Rottenberg[24] who found linearity over a considerable range of affinities. The Onsager reciprocal relations were found to hold. The observed fluxes were, however, *not homogeneous* functions of the forces as required from irreversible thermodynamics. A constant must be added to the righthand side of each flux equation. The physical meaning of these constants is not clear.[15]

Rottenberg and Gutman[25] have investigated the energy conversion combining only the first step of the respiratory chain with the phosphorylation of ADP. They used inverted spherical particles prepared from fragments of the inner mitochondrial membrane. As a result of the inversion the F_1-ATP-ase was in contact with the outer solution. This is a simplified system, which makes it easier to interpret experimental results.

The respiratory chain is reduced to the first step by poisoning the enzymes leading to the second and third steps. The inversion simplifies the phosphorylation. This reaction now takes place in the outer solution, and there is no need for the transfer processes (eqs. (10.22) and (10.23)) to replenish the reactants and remove the products of the reaction.

Rottenberg and Gutman have studied reaction velocities under varying concentration conditions. They have found that the fluxes are homogeneous linear functions of the forces in stationary state. Further information deduced from their experimental results is consistent with the assumptions that four H^+ ions are transported across the membrane for the first step of the respiratory chain and that three H^+ ions are returned across the membrane for each ADP phosphorylated.

10.2.4 *Conclusions*

Oxidative phosphorylation is the result of several coupled chemical reactions and transport processes. Important details are incompletely known. There is strong evidence, however, that the oxidation of one molecule of NADH leads to the transfer of 11 or 12 protons across the membrane, and that the production of one molecule ATP is coupled to the return of three protons across the membrane and to the return of one positive charge in the ATP–ADP exchange across the membrane. Processes transferring charges must necessarily be coupled in the stationary state, so as to give no charge accumulation. Thus the stoichiometry NADH/ATP is determined by the requirements of electroneutrality.

Until the process is better known, the application of irreversible thermodynamics is of limited value. When individual steps in the process can be studied separately, we may be able to use vectorial fluxes and forces when attacking the problem with irreversible thermodynamics. This method was used in Section 10.1 on muscular contraction. It was also used for the cell treated in Section 8.3.

10.3 *References*

1. Shimizu, T. and Johnson, K.A., *J. Biol. Chem.*, **258**, 13841 (1983).
2. Pollard, T.D. and Weihing, R.R., *CRC Crit. Rev. Biochem.*, **2**, 1 (1974).
3. Nicholls, D.G., *Bioenergetics*, Academic Press, New York, 1982.
4. Hill, A.V., *Proc. R.Soc. (B).*, **159**, 319 (1964).
5. Huxley, A.F. and Niedergerke, R., *Nature (London)*, **173**, 971 (1954).
6. Huxley, H.E. and Hanson, J., *Nature (London)*, **173**, 973 (1954).
7. Lehninger, A.L., *Principles of Biochemistry*, Worth, New York, 1982.
8. Lowe, A.G., Energetics of muscular contraction, in *Biochemical Thermodynamics* (ed. Jones, M.N.), Elsevier, Amsterdam, 1979.
9. Huxley, H.E., *Science*, **164**, 1356 (1969).
10. Cooke, R., Crowder, M.S. and Thomas, D.D., *Nature (London)*, **300**, 776 (1982).
11. Førland, T., *Biophys. J.*, **47**, 665 (1985).

12. Taylor, E.W., *CRC Crit. Rev. Biochem.*, **6**, 103 (1979).
13. Carlson, F.D. and Wilkie, D.R., Muscle Physiology, Prentice-Hall, New York, 1974.
14. Bendall, J.R., *Muscles, Molecules and Movement,* Heinemann, London, 1969.
15. Caplan, S.R. and Essig, A., *Bioenergetics and Linear Nonequilibrium Thermodynamics—The Steady State,* Harvard University Press, 1983.
16. Sleep, J.A. and Taylor, E.W., *Biochem.*, **15**, 5813 (1976).
17. Lehninger, A.L., *Ber. Bunsenges. Physik. Chem.*, **84**, 943 (1980).
18. Førland, T., Skjåk Bræk, G. and Østvold, T., *Acta Chem. Scand.*, **A33**, 647 (1979).
19. Mitchell, P., *Nature (London),* **191**, 423 (1961).
20. Padan, E. and Rottenberg, H., *Eur. J. Biochem.*, **40**, 431 (1973).
21. Nicholls, D.G., *Eur. J. Biochem.*, **50**, 305 (1974).
22. Beavis, A.D. and Lehninger, A.L., *Eur. J. Biochem.*, **158**, 315 (1986).
23. Beavis, A.D., *J. Biol. Chem.*, **262**, 6165 (1987).
24. Rottenberg, H., *Biophys. J.*, **13**, 505 (1973).
25. Rottenberg, H. and Gutman, M., *Biochem.*, **16**, 3220 (1977).

10.4 *Exercises*

10.1. *Muscular contraction—linearity of flux equation.* Use eq. (10.8) and the value $\Delta\mu_M = -32.4$ kJ mol^{-1}. Assume that the movement of a myosin head along the actin filament in one cycle in average takes place over four pairs of actin molecules. Each actin has two sites for bond formation to the myosin head, giving altogether an average of 16 steps per myosin head per cycle. (a) Estimate the average difference in chemical potential per step in units kT. (b) Estimate the deviation from the linear flux equation at maximum contraction velocity, $w = 0$. (Hint: replace $e\Delta\varphi$ in eq. (1.2) by the value calculated in (a)).

10.2. *Energy conversion in mitochondria.* The change in Gibbs energy by oxidation of NADH, ΔG_{ox}, is assumed to be -210 kJ mol^{-1}. Assume further that 8% of the Gibbs energy is dissipated as heat in the process of transferring H^+ from the inside to the outside of the membrane. The Gibbs energy change for the phosphorylation, ΔG_{ph}, is assumed to be 67 kJ.mol^{-1}. (a) Calculate the difference in electrochemical potential for H^+, $\Delta\tilde{\mu}_{H^+} = \tilde{\mu}_{H^+}(\text{in}) - \tilde{\mu}_{H^+}(\text{out})$ (units mV), when the number of H^+ transferred across the membrane per NADH oxidized, n, is equal to (i) 12, (ii) 9, (iii) 6. (b) Assume that the number of H^+ transferred from outside to inside per ATP synthesized, m, is (i) 2, (ii) 3, (iii) 4. Calculate $\Delta\tilde{\mu}_{H^+}$ for each case. (c) What combinations of n and m permit the synthesis of ATP?

10.3. *Electrochemical potential difference.* The difference in electrochemical potential for H^+ across a membrane, $\Delta\tilde{\mu}_{H^+}$, can be defined operationally as the emf of a cell with hydrogen electrodes on both sides of the membrane, $\Delta\varphi'$ (cf. eq. (4.39), $\Delta\tilde{\mu}_{H^+} = \Delta\varphi'$). (a) Consider the cell

$$(\text{Pt})\text{H}_2(\text{g},1\ \text{atm})|\text{HX}(\text{aq},c_{1,\text{I}}),\text{KX}(\text{aq},c_{2,\text{I}})|\text{Membrane}|$$
$$\text{HX}(\text{aq},\ c_{1,\text{II}}),\ \text{KX}(\text{aq},c_{2\ \text{II}})|\text{H}_2(\text{g},1\ \text{atm})(\text{Pt}).$$

The membrane is permeable to the neutral molecule HX and to K^+ ions, but impermeable to KX and to H^+. The conditions are similar to those of the inner mitochondrial membrane when valinomycin has been added. Assume equilibrium for the distribution of HX across the membrane, $\mu_{HX,I} = \mu_{HX,II}$. The emf of the cell is $\Delta\varphi'$. Show that $\Delta\varphi' = -\Delta\mu_{KX} = \mu_{KX,II} - \mu_{KX,I}$. (b) Use eq. (10.25) and show that $\Delta\tilde{\mu}_{H^+} = -\Delta\mu_{KX}$ for an ideal solution, where $c_{H^+,I}\,c_{X^-,I} = c_{H^+,II}\,c_{X^-,II}$.

CHAPTER 11

Non-isothermal Transport

In non-isothermal systems the transport of mass and charge is affected by the temperature differences in the system. Different types of systems with temperature differences were treated in Chapter 6. The effect of temperature differences on mass transport in a homogeneous system, the Soret effect, is usually small. In an inhomogeneous system the thermal osmosis may give a large effect (see Section 6.2). The coupling of charge transport and heat transport is small in an all-metal circuit. In an electrochemical cell, however, the coupling may be substantial (see Section 6.6.1).

The coupling between fluxes was discussed in Section 3.4. When two fluxes are linearly dependent, $J_i = \alpha J_j$, the coupling coefficient, α, is equal to a ratio between coefficients, $\alpha = l_{ij}/l_{jj}$ or $\alpha = L_{ij}/L_{jj}$. When there is no coupling, the cross coefficient is equal to zero.

In non-isothermal systems the set of transport equations includes J_q, the measurable heat flux. As was explained in Section 2.1.2 measurable heat transferred over a temperature difference is not constant and thus neither is J_q constant over distance when there is a temperature gradient, not even under stationary conditions.

In this chapter we shall study two applications of irreversible thermodynamics to non-isothermal systems where transport equations include J_q. In the first case we shall examine the coupled transport of heat and mass in frost heave. In the second case we shall examine the coupled transport of heat, mass and charge in electrochemical cells with molten NaOH as the electrolyte. Experimental and theoretical results are compared.

11.1 *Frost Heave*

When a soil containing water freezes, its surface heaves, often unevenly; this is called frost heave. When the soil thaws again, it becomes soft and muddy. These phenomena are serious problems in construction engineering. In areas where frost heave occurs, one must take extra care with the foundation of buildings and roads.

Frost heave can be a large effect where the average temperature of the coldest month is below 0°C. The geographical region where frost heave occurs is extensive, covering large parts of North America, Europe and Asia. The extent of the frost heave also depends on the soil. There is very little frost heave in coarse sandy soils, while there is a much larger effect in fine grain soils like clay, and particularly silt.

When a soil has been allowed to settle, the solid particles are packed together so that they are in contact with one another. The solid particles in contact form a framework which is able to resist pressure. There will be a distribution of pore sizes between the particles. These pores may be more or less filled with water. The water in these pores is called *bound water*. If there is more water in the soil than the space between the particles in contact permits, the framework must expand. Contact between particles is lost and the soil loses its strength. The excess of water is called *free water*.

In frost heave the soil expands more than the expansion due to the freezing of the bound water in the unfrozen soil. Excess of water forms ice lenses under the surface. When the soil thaws from the top, the excess of free water makes it soft. This excess of water was *transported* to the ice lenses during freezing.

11.1.1 *Transports during frost heave*

In order to study the transports more closely, we shall look at the frost heave in more detail. Figure 11.1 gives a schematic picture of a vertical column of soil where frost heave is taking place.

The transports of water and heat will be considered in two steps. The capillary rise of water from the table of water, level 1, through the unfrozen soil at temperatures above freezing, up to level 2 at the freezing point is the first step. The transport of heat through this region will also be considered. The second step is the transport of water and heat from level 2 through the partly frozen soil to the ice lens at level 3. Here thermal osmosis takes place, in addition to capillary rise.

First step. The pores of the fine-grain soil act as a set of capillaries. These are filled by capillary rise from the table of

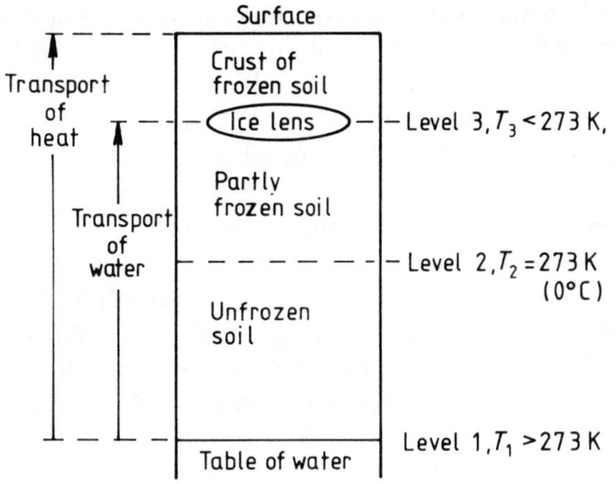

Figure 11.1. A section of freezing soil. Schematic picture of frost heave.

water. As we shall see when discussing the second step, water is transported away from level 2. If there is no access to air, the void must be filled with water flowing through the soil from the table of water. The force for this flux is given by a pressure reduction at level 2. The temperature difference between levels 1 and 2 will cause a flux of heat.

At level 2 there is an equilibrium between ice in the pores and liquid water. The relative amounts of the two phases can change without a change in Gibbs energy. We include in the first step the freezing of all the transported water at level 2. The water is transported in liquid form in the second step, and this means that the second step will involve the remelting of the ice at level 2.

Second step. The ice melts at level 2. Liquid water is transported through pores of the partly frozen soil in the direction of decreasing temperature. The temperature is below 0°C. Liquid water below 0°C is a well-known phenomenon in capillaries. In the smallest pores the water will remain liquid at temperatures several degrees below 0°C, while ice can form at 0°C and atmospheric pressure in the largest pores. At a sufficiently low temperature even the water in the smallest pores will freeze. Below this temperature there is a solid crust of ice which does not permit any transport of water. The transport process from level 2 to level 3 can be treated as thermal osmosis (see Section 6.2). Level 2 can be compared to the heat reservoir I illustrated in Fig. 6.2. The ice melts absorbing heat. The partly frozen soil between levels 2 and 3 may be considered as the membrane. Level 3 corresponds to heat reservoir II. Here the water freezes and heat is released. Under a solid crust of ice the thermal osmosis leads to an increased pressure at level 3. The thermal osmosis is a slow process.

The increased pressure under the crust acts on the solid framework of the soil. When an ice crystal at level 3 has grown to fill the pore, it will be subject to the same pressure. The wall of the pore breaks at a weak point, and the crystal will continue to grow. In the end a lens of ice is formed below the frozen crust. As the ice lens grows, the pressure under the frozen crust increases until the transport of water stops, or until the crust breaks. When the surface temperature decreases, the frozen crust grows in thickness. The border between the crust of frozen soil and the partly frozen soil (level 3) is pushed further down in the ground, and a new lens may form at this lower level.

11.1.2 *Flux equations for the transports*

There is experimental evidence that there is little or no coupling between the flux of water and the measurable heat flux in the *first step*. This will be discussed in Section 11.1.3. The heat flux is given by the thermal conductivity and the temperature difference, $J_q = -\lambda\Delta T$. The flux of water is given by a transport coefficient, $\bar{l}$, and the difference in chemical potential for water. With pure water (or no variation in concentration of impurities) the difference in chemical potential is given by $\Delta\mu_{H_2O} = V_{H_2O}\Delta p$ (see eq. (5.16)). Thus we have $J_{H_2O} = -\bar{l}\,V_{H_2O}\Delta p$. The pressure difference is the pressure reduction created at level 2 by the thermal osmosis in the second step. Heat is released when water freezes at level 2.

The *second step* is treated as thermal osmosis, and the fluxes can be expressed by equations similar to eqs. (6.4):

$$J_q = -\bar{l}_{11}\Delta\ln T - \bar{l}_{12}\Delta\mu_{\text{ice},T} \tag{11.1a}$$

$$J_w = -\bar{l}_{21}\Delta\ln T - \bar{l}_{22}\Delta\mu_{\text{ice},T} \tag{11.1b}$$

We have $\Delta\ln T = \ln(T_3/T_2)$, and with pure ice at both levels we have

$$\Delta\mu_{\text{ice},T} = V_{\text{ice}}\Delta p \tag{11.2}$$

where V_{ice} is the molar volume of ice and Δp is the difference between the pressure acting on the ice lens at level 3 and on an ice crystal at level 2. Ice and liquid water are in equilibrium at level 2 and the pressure on the ice is the same as the pressure on the water in the pores.

The maximum possible frost heave pressure is the pressure that completely suppresses the flux of water, $J_w = 0$. This pressure is found from eqs. (11.1b) and (11.2) (cf. eq. (6.13):

$$\Delta p_{J_w=0} = -(1/V_{ice})(\bar{l}_{21}/\bar{l}_{22})\Delta \ln T \tag{11.3}$$

Further, since $\bar{l}_{12} = \bar{l}_{21}$, we have, similarly to eq. (6.14)

$$(J_q/J_w)_{\Delta T=0} = \bar{l}_{12}/\bar{l}_{22} = q_w^* \tag{11.4}$$

where q_w^* is the *heat of transfer*. The transfer of *liquid* water between levels 2 and 3 gives only negligible contribution to the transfer of heat. This is in agreement with the lack of coupling between J_q and J_w in the first step. The main contribution to the heat of transfer is given by the melting of ice at level 2 with absorption of heat and the freezing of the water at level 3 with release of heat.

In contact with a solid surface there is always a layer of water with properties different from those of bulk water. This layer stays unaltered, however, by the transport process, and the phase changes are between ice and *mobile* water with bulk water properties. The heat of transfer is equal to the enthalpy of melting:

$$q_w^* = \Delta H_m \tag{11.5}$$

The maximum frost heave pressure can be expressed in terms of the enthalpy of melting. From eqs. (11.3)–(11.5) we have

$$\Delta p_{J_w=0} = -(\Delta H_m/V_{ice})\Delta \ln T \tag{11.6}$$

The enthalpy of melting varies with pressure and temperature. The variation with pressure is negligible within the pressure ranges of frost heave, while the variation with temperature may be of importance when there is a large temperature difference between levels 2 and 3. The variation of V_{ice} with pressure and temperature is negligible.

When ΔT is much smaller than T, $\Delta \ln T \approx \Delta T/T$. Since $\Delta H_m/T_m = \Delta S_m$, we obtain

$$\Delta p_{J_w=0} = -(\Delta S_m/V_{ice})\Delta T \tag{11.7}$$

At 0°C and 1 atm pressure $\Delta S_m = 22.0$ J K^{-1} mol^{-1} and $V_{ice} = 1.96 \times 10^{-5}$ m^3 mol^{-1}. When the temperature of the ice is not too low, we thus have

$$\Delta p_{J_w=0} = -11.2 \times 10^5 \, \Delta T \text{ Pa} \tag{11.8}$$

$$(\text{or } -11.1\, \Delta T \text{ atm or } -11.4\, \Delta T \text{ kg cm}^{-2})$$

If a soil permits transport of water down to a temperature of −5°C, an ice lens will grow at this temperature. The flux of water will not be suppressed until a pressure of more than 50 atm has been built up.

In eq. (11.1b) $\bar{l}_{22}\Delta H_m$ may be substituted for $\bar{l}_{21}$ (see eqs. (11.4) and (11.5)), and further $\Delta S_m \Delta T$ for $\Delta H_m \Delta \ln T$. We may also substitute $V_{ice}\Delta p$ for $\Delta \mu_{ice}$ (see eq. (11.2)):

$$J_w = -\bar{l}_{22}(\Delta S_m \Delta T + V_{ice}\Delta p) \tag{11.9}$$

The flux of water, J_w, is equal to the rate of growth for the ice lens, and $\bar{l}_{22}$ is the average hydraulic permeability for the total transport path for water. The value of $\bar{l}_{22}$ may chiefly be determined by low local values of l_{22} close to the ice lens. The value may also decrease with time as the ratio of ice to liquid water increases in the partly frozen soil.

11.1.3 *Experimental studies of frost heave*

Frost heave has been investigated in laboratories under controlled conditions. We shall refer to measurements carried out by Takashi and coworkers.[1] Their experimental arrangement is shown schematically in Fig. 11.2. To emphasize the resemblance to frost heave in nature, the apparatus has been rotated through 180° in Fig. 11.2. In the actual experiment the cooling plate was at the bottom and the piston at the top.

The soil sample is contained in a transparent, thermally insulated cylinder. In preparation for the experiment the sample is subjected to a high pressure in order to obtain a good packing of the soil particles to form a strong framework. The pressure is released and the pore water is de-aerated. Thereafter partial freezing of the sample is achieved by keeping the temperature at both ends constant, the piston at a temperature above 0°C and the cooling plate at a temperature below 0°C. During this freezing there is free access to a reservoir of water.

When the sample has been prepared in this way, the valve for the pore water supply is closed, and the sample is subjected to a pressure by means of the piston. The pressure is taken up by the framework of the soil. The pore water will also be under increased pressure, but lower than the pressure on the framework. Both pressure on the piston, p_2, and in the pore water, p_1, are measured.

When the difference in pressure, $\Delta p = p_2 - p_1$, is less than the maximum frost heave pressure obtainable for the chosen temperature on the cooling plate, there will be a flux of water towards the cooling plate, $J_w > 0$, the ice lens will grow and the

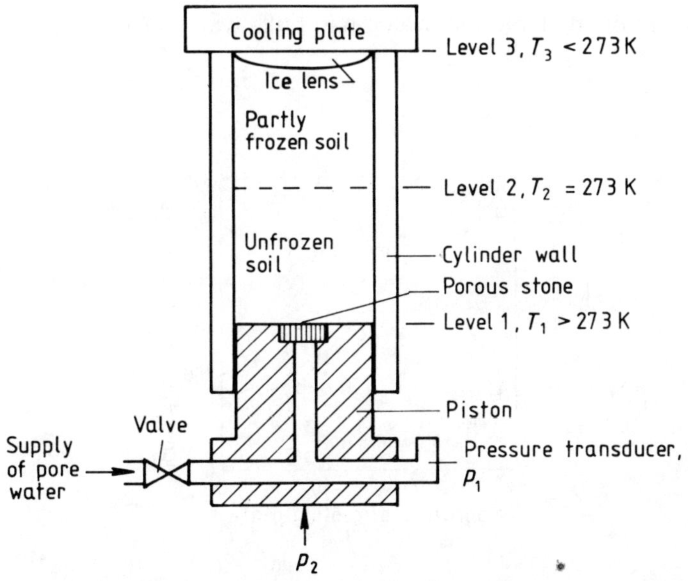

Figure 11.2. Schematic picture of experimental arrangement for measuring frost heave. Apparatus rotated through 180°.

pressure difference will increase over time. When the difference is larger than the maximum frost heave pressure, the flux of water will be in the opposite direction, $J_w < 0$, the ice lens will be melting and the pressure difference will decrease over time. Thus one can approach the maximum frost heave pressure from both sides.

Takashi and coworkers tested two types of soil, Mañaita-bridge clay and Negishi silt. The time needed to obtain maximum frost heave pressure was long, ranging from 500 to 2000 hours. They measured the upper limit of frost heave pressure, σ_u(kg cm^{-2}), for different temperatures on the cooling plate, θ_c(°C). The pressure σ_u corresponds to $\Delta p_{J_w=0}$. Assuming that the first ice crystals in pores form at 0°C, we have $\theta_c = T_3 - T_2 = \Delta T$. In the temperature range 0 to −17°C for the clay, and 0 to −4°C for the silt, the linear relation was found, $\sigma_u = -11.4\ \theta_c$. This agrees well with the theoretical value given in eq. (11.8). The experimental results are given in Figs. 11.3 and 11.4.

It can be seen from Figs. 11.3 and 11.4 that when θ_c approaches 0°C, the σ_u approaches zero pressure difference within the limits of experimental error. As we shall see below, this observation makes possible a check of eq. (11.5), which is based on the assumption that mobile water has bulk water properties. It can

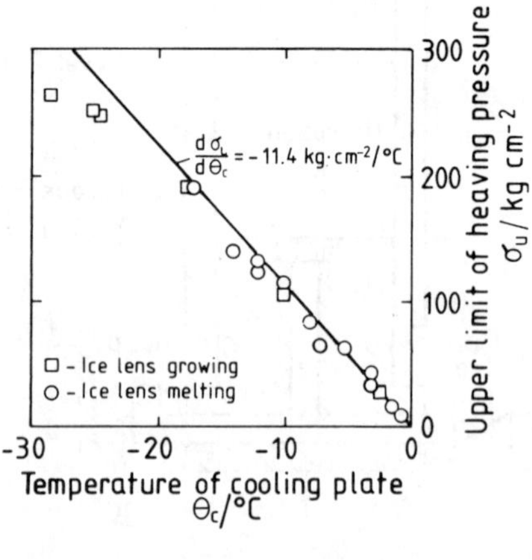

Figure 11.3. Frost heave in Manaita-bridge clay. Reproduced by permission of the 2nd International Symposium on Ground Freezing.[1]

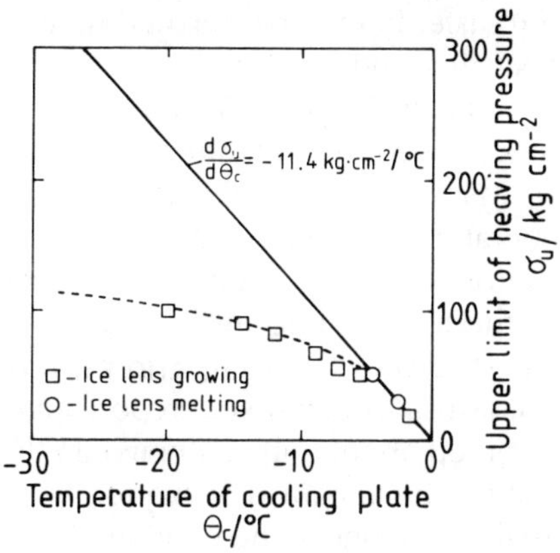

Figure 11.4. Frost heave in Negishi silt. Reproduced by permission of the 2nd International Symposium on Ground Freezing.[1]

also be deduced that there is no coupling between transports of heat and water in the first step.

In most experiments carried out by Takashi and coworkers, the temperature of the piston was + 6°C. For $\theta_c = 0$°C, the experimental conditions can be described by the following cell:

H_2O(l), 6°C|Porous stone|Water in soil||Ice lens, 0°C

In this case there is no pressure difference and the temperature difference does not cause a transport of water. The ice lens is in equilibrium with liquid water at the same temperature. If the ice lens were replaced by liquid water, there would still be no transport of water. The following arrangement:

H_2O(l), 6°C|Porous stone|Water in soil|H_2O(l), 0°C

would have given some thermal osmosis effect and shown a pressure increase on the low-temperature side, if there were a significant heat of adsorption for the mobile water in the soil. Since there is no pressure increase, the enthalpy of mobile water in the soil is approximately equal to the enthalpy of bulk water and eq. (11.5) is valid. With no significant heat of adsorption, there is no heat of transfer in the first step and no coupling between the measurable heat flux and the flux of water.

The results illustrated in Figs. 11.3 and 11.4 show deviations from the straight line at lower temperatures. The deviation is small for the clay and it may, at least partly, be explained by changes in ΔS_m with temperature. At -20°C ΔS_m is about 13% smaller than at 0°C. Equation (11.7) was derived assuming a constant value at ΔS_m over the temperature interval ΔT.

The deviation for the silt is too large to be explained by the above changes. Takashi and coworkers report that in some cases an ice lens was observed about midway between the freezing front and the cooling plate. This means that the ice lens is formed at a higher temperature than the one at the cooling plate, corresponding to a lower maximum frost heave pressure. We can also see from Fig. 11.4 that measurements at the lower temperatures have only been taken with a growing ice lens; no measurements with a melting ice lens are reported. As the rate of growth is very low at the low temperatures, there is a possibility that maximum frost heave pressure was not obtained.

11.1.4 *Conclusions*

The important parameters for evaluation frost heave are given in eq. (11.9), which expresses the rate of growth of an ice lens. The parameter $\overline{l}_{22}$, the average hydraulic permeability, depends on the type of soil, the distances between the water table, the freezing front where the first crystals of ice appear and the ice lens. The second important parameter, ΔT, is mainly given by the temperature (°C) of the lowest lying ice lens. The third important parameter is the difference in pressure, Δp. As the pressure difference increases with time, the rate of growth of the ice lens decreases. If the crust of frozen soil is not able to withstand the increased pressure, and breaks, the flux of water may continue indefinitely as long as temperature is low enough to allow ice formation.

Frost heave can be prevented if the continuity of water between the water table and the freezing front is broken. It is known empirically that a layer of coarse sand or gravel deep in the soil and ditches to drain the water reduce frost heave.

11.2 *Peltier Effects in a Thermocell*

The heat balances are important factors in the industrial use of electrochemical cells. Uncontrolled temperature changes in a cell, or in a part of the cell, may be disastrous. Overheating may lead to, for example, breakdown of cell materials or boiling of the electrolyte, while undercooling may lead to freezing of the electrolyte or other calamities. Apart from these more drastic effects, the heat balances are of importance for the economy of the cell process.

In order to obtain a detailed picture of temperatures in the cell, it is not sufficient to know the total heat of cell reaction and the heat supplied to the cell (or removed). We must also have knowledge about heat absorbed and heat evolved at the individual electrodes, the *Peltier effects* at the electrodes. There are other heat effects to be considered, heat due to overpotential at electrodes and due to ohmic resistance. These effects will not be discussed here, although they belong in an evaluation of the heat balance. This section will deal with the Peltier effect.

The Peltier effect in thermocells was discussed in Section 6.6. In this section the theories will be applied to the electrolysis of water in molten sodium hydroxide with platinum electrodes. The heat balance of a cell of this kind has been studied by Ito and coworkers,[2–4] who have found substantial Peltier effects at the electrodes. Their cell operates at temperatures around 350°C. One advantage of the cell is very low overpotentials on the electrodes.

11.2.1 *Electrolysis of water in molten sodium hydroxide. Peltier effects*

We shall study the electrolysis of water in the following cell:

$$(Pt)O_2(g),H_2O(g)|NaOH(l)|H_2O(g),H_2(g)\ (Pt) \qquad (11.10)$$

The cathode reaction is $H_2O(g) + e^- = \frac{1}{2} H_2(g) + OH^-$. The OH^- produced migrates through the molten sodium hydroxide to the anode where the anode reaction is $OH^- = e^- + \frac{1}{4} O_2(g) + \frac{1}{2} H_2O(g)$. When one faraday of positive charge passes from left to right through the inner circuit of the cell, one half mole $H_2(g)$ is gained at the cathode, while one mole $H_2O(g)$ is lost. At the anode one fourth mole $O_2(g)$ and one half mole $H_2O(g)$ is gained. The total reaction is,

$$\tfrac{1}{2}\,H_2O(g) = \tfrac{1}{2}\,H_2(g) + \tfrac{1}{4}\,O_2(g) \tag{11.11}$$

Because of the chemical reactions at the electrodes, substantial Peltier effects can be expected.

The above cell cannot be directly compared to the cells treated in Section 6.6, since only cells with identical cathode and anode were treated there. In order to find the Peltier heats in cell (11.10) we may study the following two cells:

$$(T)\ (Pt)O_2(g),H_2O(g)|NaOH(l)|H_2O(g),O_2(g)\ (Pt)\ (T + \Delta T) \tag{12.12}$$

and

$$(T)\ (Pt)H_2(g),H_2O(g)|NaOH(l)|H_2O(g),H_2(g)\ (Pt)\ (T + \Delta T) \tag{11.13}$$

The chemical potentials of the three species, H_2O, H_2 and O_2, are related by an equilibrium equation, and only two of these can be considered as components.

Let us first consider cell (11.12) with identical electrodes, $\Delta\mu_{H_2O,T} = 0$ and $\Delta\mu_{O_2,T} = 0$. There are then only two forces in the system, $-\Delta\ln T$ and $-\Delta\varphi$.

The dissipated energy per time unit is

$$\frac{TdS}{dt} = -\Delta\ln T\,J_q - \Delta\varphi I \tag{11.14}$$

The corresponding flux equations are

$$J_q = -\bar{L}_{11}\Delta\ln T - \bar{L}_{12}\Delta\varphi \tag{11.15a}$$

$$I = -\bar{L}_{21}\Delta\ln T - \bar{L}_{22}\Delta\varphi \tag{11.15b}$$

By inspection it is found that the equations are quite similar to eqs. (6.33) and (6.34), and that cell (11.12) can be treated in the same way as cell a of Section 6.6.1. We shall adopt the same marking of the symbols, thus the electric potential is named $\Delta\varphi_a$ and the Peltier heat is named $\pi_{T,a}$.

The Peltier heat is defined by eq. (6.35): $(J_q/I)_{\Delta T=0} = \bar{L}_{12}/\bar{L}_{22} = \pi_{T,a}$. Experimental values for the Peltier heat can be found by recording corresponding values for $\Delta\varphi_a$ and ΔT for $I \approx 0$, since they are related by eq. (6.37), $\Delta\varphi_a = -\pi_{T,a}\Delta \ln T \approx -(\pi_{Ta}/T)\Delta T$. Here $\pi_{T,a}/T$ is the entropy transferred. As an approximation we may assume the variation in this entropy with temperature to be negligible. This is the same as assuming a negligible Thomson heat. Then we have $\pi_T/T = \pi_{(T+\Delta T)}/(T+\Delta T) = -\Delta\varphi/\Delta T$ (cf. eq. (6.24)).

We shall consider the entropy balance across the electrode–electrolyte interface for the left-hand-side electrode when one faraday of positive charge passes from left to right through the inner circuit of the cell. As was discussed in Section 6.6.1, the reversible entropy balance is independent of the frame of reference. For convenience NaOH(l) may be chosen as the frame of reference. The interface receives entropy from the heat reservoir, $\pi_{T,a}/T$, entropy transported through the electrode to the interface, S^*_{Pt}, and entropy transported through the electrolyte to the interface, $t_{OH^-}S^*_{OH^-}$. With NaOH(l) as frame of reference, the transference number of OH^-, $t_{OH^-} = 1$. The production of one fourth mole O_2 consumes the entropy $\frac{1}{4} S_{O_2}$, and the production of one half mole H_2O consumes the entropy $\frac{1}{2} S_{H_2O}$. Hence we have

$$\pi_{T,a}/T + S^*_{Pt} + S^*_{OH^-} = \tfrac{1}{4} S_{O_2} + \tfrac{1}{2} S_{H_2O}$$

or

$$\pi_{T,a}/T = -(S^*_{Pt} + S^*_{OH^-}) + \tfrac{1}{4} S_{O_2} + \tfrac{1}{2} S_{H_2O} \tag{11.16}$$

The cell (11.13) can be treated in a similar way. It can be compared with cell b of Section 6.6.2. We shall adopt the same marking of the symbols, naming the electric potential $\Delta\varphi_b$ and the Peltier heat $\pi_{T,b}$. The relation between these two quantities when $I \approx 0$ is given by eq. (6.43), $\Delta\varphi_b = -\pi_{T,b}\Delta \ln T \approx -(\pi_{T,b}/T)\Delta T$.

The entropy balance across the electrode–electrolyte interface for the left-hand-side electrode upon the passage of one faraday of positive charge from left to right through the inner circuit of cell (11.13), is studied in a similar way as for cell (11.12). The interface receives entropy from the heat reservoir, $\pi_{T,b}/T$, the entropy transported through the electrode to the interface, S^*_{Pt}, and the entropy transported through the electrolyte to the interface, $S^*_{OH^-}$. Again, with NaOH(l) as the frame of reference, we have $t_{OH^-} = 1$. The disappearance of one half mole H_2 liberates the entropy $\frac{1}{2}S_{H_2}$, while the production of one mole H_2O consumes the entropy S_{H_2O}. Thus we have

$$\pi_{T,b}/T + S^*_{Pt} + S^*_{OH^-} + \tfrac{1}{2} S_{H_2} = S_{H_2O}$$

or

$$\pi_{T,b}/T = -(S^*_{Pt} + S^*_{OH^-}) - \tfrac{1}{2} S_{H_2} + S_{H_2O} \qquad (11.17)$$

Remembering that $\Delta\varphi/\Delta T = -\pi_T/T$ we have

$$\Delta\varphi_a/\Delta T - \Delta\varphi_b/\Delta T = \tfrac{1}{2} S_{H_2O} - \tfrac{1}{2} S_{H_2} - \tfrac{1}{4} S_{O_2} = -\Delta S \qquad (11.18)$$

where ΔS is the entropy change for the reaction given by eq. (11.11), the cell reaction of the isothermal cell (11.10).

11.2.2 *Experimental determinations of Peltier heats*

The Peltier heats at both electrodes of cell (11.10) was studied by Ito and coworkers.[5] A Na|Na$^+$ electrode at constant temperature was used as a reference electrode combined with each half cell at different temperatures. Corresponding values for $\Delta\varphi_a$ and $\Delta T = T_2 - T_1$ were obtained by the following measurements:

(T_1) Ref. electrode|NaOH(l)|H_2O(g),O_2(g) (Pt) (T_1) $\quad\Delta\varphi_1$

(T_1) Ref. electrode|NaOH(l)|H_2O(g),O_2(g) (Pt) (T_2) $\quad\Delta\varphi_2$

(11.19)

For cell (11.12) with identical electrodes, $\Delta\varphi_a = \Delta\varphi_2 - \Delta\varphi_1$ when $T = T_1$ and $T + \Delta T = T_2$. In a similar way corresponding values for $\Delta\varphi_b$ and ΔT were obtained.

Experimental values for the emf of cell (11.19), $\Delta\varphi_2/F$, when $p_{H_2O} = 0.2$ atm for $p_{O_2} = 0.8$ atm are shown in Fig 11.5. The experimental values for the emf of cell (11.20), when $p_{H_2O} = 0.2$ atm and $p_{H_2} = 0.8$ atm are shown in Fig. 11.6.

$$(T_1)\ \text{Ref. electrode}|\ \text{NaOH(l)}|\text{H}_2\text{O(g)},\text{H}_2\text{(g) (Pt)}\ (T_2) \qquad (11.20)$$

Both figures show a linear relation between emf and $(T_2 - T_1)$. This confirms that the Thomson heat is negligible, i.e. that $\pi_T/T = \pi_{(T+\Delta T)}/(T+\Delta T)$. From the slopes of the lines the following experimental values are obtained:

$$\Delta\varphi_a/\Delta T = -\ 115 \pm 4 \quad \text{J K}^{-1}\ \text{faraday}^{-1} \qquad (11.21)$$

and

$$\Delta\varphi_b/\Delta T = -\ 101 \pm 2 \quad \text{J K}^{-1}\ \text{faraday}^{-1} \qquad (11.22)$$

The difference between the experimental values is (cf. eq. (11.18))

$$\Delta\varphi_a/\Delta T - \Delta\varphi_b/\Delta T = -\ \Delta S = -\ 14 \pm 6 \quad \text{J K}^{-1}\ \text{faraday}^{-1} \qquad (11.23)$$

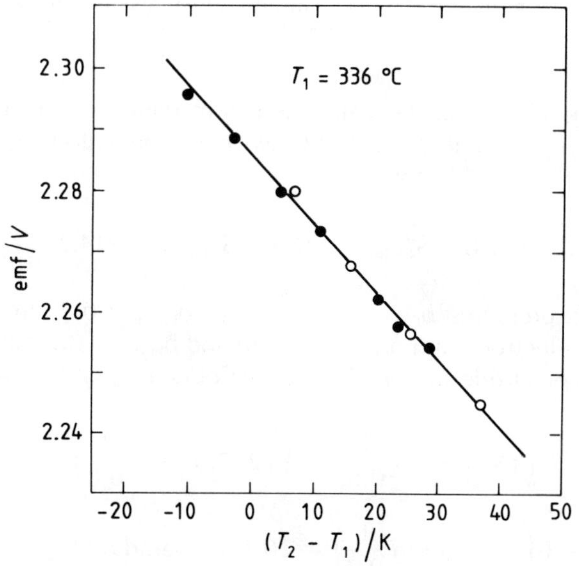

Figure 11.5. Electromotive force of cell (11.19) (see text). ○: Measured when T_2 was increased; •: measured when T_2 was decreased. Reprinted by permission of the publisher, The Electrochemical Society, Inc., from Ref. 5.

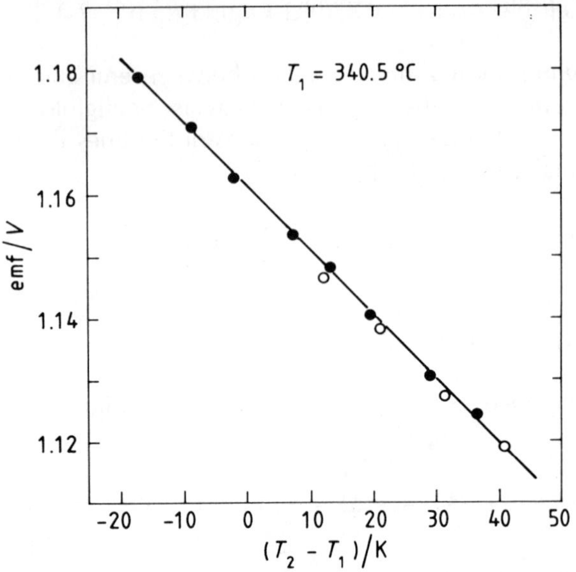

Figure 11.6. Electromotive force of cell (11.20), (see text). ○: Measured when T_2 was increased; ●: measured when T_2 was decreased. Reprinted by permission of the publisher, The Electrochemical Society, Inc., from Ref. 5.

The value of ΔS can be calculated from data given in tables of thermodynamic data.[6] At 350°C we have the following entropy values (all in J K^{-1} mol^{-1}):

$$S^o_{H_2(g)} = 151.8, \quad S^o_{O_2(g)} = 227.5, \quad S^o_{H_2O(g)} = 214.2$$

With the pressures $p_{O_2} = 0.8$ atm and $p_{H_2O} = 0.2$ atm for the (Pt)O_2(g) electrode and $p_{H_2} = 0.8$ atm and $p_{H_2O} = 0.2$ atm for the (Pt)H_2(g) electrode in cell (11.10), we obtain

$$\Delta S = \tfrac{1}{2}\left(151.8 + R\ln\frac{1}{0.8}\right) + \tfrac{1}{4}\left(227.5 + R\ln\frac{1}{0.8}\right) - \tfrac{1}{2}\left(214.2 + R\ln\frac{1}{0.2}\right) = 20 \text{ J K}^{-1}\text{ faraday}^{-1} \qquad (11.24)$$

The value agrees with the experimental value within the limits of error.

The Peltier heat for each electrode in cell (11.10) can be obtained from the experimental values of $\Delta\varphi/\Delta T$. At 350°C (623 K) we have

$$\pi_a = -(\Delta\varphi_a/\Delta T)T = 115\cdot623\cdot10^{-3} = 72 \text{ kJ faraday}^{-1}$$

$$-\pi_b = +(\Delta\varphi_b/\Delta T)T = -101\cdot623\cdot10^{-3} = -63 \text{ kJ faraday}^{-1}$$

For each faraday of charge passing through the cell, 72 kJ are absorbed at the anode, while 63 kJ are released at the cathode. Unless compensated for, this leads to cooling of the anode and heating of the cathode.

Water dissolves in liquid sodium hydroxide and the enthalpy of solution is -60 kJ mol^{-1}. A cell may be constructed in such a way that water vapour is dissolved in the electrolyte at some distance from the cathode. Ito and coworkers have designed a cell with this feature.[4] In such a cell the cathode reaction will be $H_2O(NaOH) + e^- = \frac{1}{2} H_2(g) + OH^-$, where $H_2O(NaOH)$ is water dissolved in NaOH. The dissolved water has a lower entropy than the water vapour, and the term S_{H_2O} of eq. (11.17) has a lower value. The Peltier heat $\pi_{T,b}$ is smaller and less heat is released at the cathode. Instead heat is released where the dissolution process for the water takes place.

The Peltier heats are not changed by $I > 0$. When the cell is working, however, gas bubbles of H_2 and O_2 will ascend through the electrolyte at the cathode and anode, respectively. This will cause some vaporization of dissolved water, and thus contribute to a change in the heat effects at the electrodes. Ito and coworkers have given a detailed calculation of the energy balance for the cell of their design.[4]

11.2.3 *Entropy transported with ions in molten sodium hydroxide*

The entropy transported with one mole OH^- ions in molten sodium hydroxide is $S^*_{OH^-}$. Its value can be obtained from eq. (11.6) or eq. (11.17) when the value is known for all the other terms in the equation.

In addition to the entropies for H_2O, H_2 and O_2 and the experimental values for Peltier heats, a value is needed for S^*_{Pt}. The transported entropy for electrons in platinum is about

1 J K^{-1} faraday^{-1} at 350°C.[7] Since S^*_{Pt} refers to positive charge, we have

$$S^*_{Pt} = -1 \text{ J K}^{-1} \text{ faraday}^{-1}$$

The average value of $S^*_{OH^-}$ obtained from eqs. (11.16) and (11.17) is

$$S^*_{OH^-} = 54 \text{ J K}^{-1} \text{ faraday}^{-1} \quad \text{at } 350°\text{C} \tag{11.25}$$

According to eq. (6.40) we have

$$S_{NaOH(l)} = S^*_{Na^+} + S^*_{OH^-} \tag{11.26}$$

The value of $S_{NaOH(l)}$ at 350°C is 138 J K^{-1} mol^{-1},[6] and thus we have

$$S^*_{Na^+} = 84 \text{ J K}^{-1} \text{ faraday}^{-1} \quad \text{at } 350°\text{C} \tag{11.27}$$

11.2.4 *Conclusions*

The example treated here shows that the Peltier heats of the electrodes may be substantial, and much larger than the reversible heat of reaction for the cell. A large contribution to the Peltier heat of an electrode comes from the reactants disappearing and products being formed at the electrodes. For known half-reactions one can easily estimate this contribution from known values of the entropies of the species. The transported entropy may also give a considerable contribution to the Peltier heat. Information about transported entropies, however, is scarce and in many cases the total value of a Peltier heat can only be determined experimentally.

Measurements of the thermoelectric power show that the Thomson heat is negligible in molten sodium hydroxide.

11.3 *References*

1. Takashi, T., Ohrai, T., Yamamoto, H. and Okamoto, J., *The 2nd International Symposium on Ground Freezing, 1980,* Norw. Inst. Technology, Trondheim, 713.
2. Ito, Y., Foulkes, F.R. and Yoshizawa, S., *J. Electrochem. Soc.,* **129**, 1936 (1982).
3. Ito, Y., Hayashi, H., Hayafuji, N. and Yoshizawa, S., *Electrochimica Acta,* **30**, 701 (1985).
4. Ito, Y., Kaiya, H. and Yoshizawa, S., *Energy Dev. Jpn,* **3**, 153 (1981).
5. Ito, Y., Kaiya, H, Yoshizawa, S., Ratkje, S.K. and Førland, T., *J. Electrochem. Soc.,* **131**, 2504 (1984).
6. Barin, I. and Knacke, O. (eds), *Thermochemical Properties of Inorganic Substances*, Springer Verlag, Berlin 1973.
7. Moore, J.P. and Graves, R.S., *J. Appl. Phys.,* **44**, 1174 (1973).

11.4 *Exercises*

11.1. *Change of frame of reference.* In Section 11.2.1 NaOH(l) was used as the frame of reference for the cell (eq. (11.12)) when the reversible entropy balance across the electrode–electrolyte interface was studied. This frame of reference implies that $t_{Na^+} = 0$ in the electrolyte. We shall choose the walls of the cell as the frame of reference. Let us assume that the electrolyte, NaOH(l), moves relatively to the walls of the cell giving finite transference numbers for both Na^+ and OH^- measured against the walls of the cell. (a) Derive an equation for the thermoelectric power, $\Delta\varphi/\Delta T$, using the walls of the cell as the frame of reference. (b) Show that this thermoelectric power is identical to the one obtained with NaOH(l) as the frame of reference. (Hint: use eq. (6.40)).

Appendices

APPENDIX 1

Thermodynamic Equations

Partial differentiation

Let Y be a function of several variables $Y = f(x_1, x_2, \ldots)$

Partial differentiation, total differential $dY = \Sigma_i \left(\frac{\partial Y}{\partial x_i}\right)_{x_{j \neq i}} dx_i$

Fundamental thermodynamic functions

First law $dU = dq + dw \quad \oint dU = 0$
Second law $dS = dq_{rev}/T$
Third law $S_{T=0} = 0$
Relations between thermodynamic functions:

Internal energy	U
Enthalpy	$H = U + pV$
Gibbs (free) energy	$G = H - TS$
Helmholtz energy	$A = U - TS$

Thermodynamic functions for closed systems

Internal energy of a closed system expressed by internal, extensive variables $U = f(S,V)$

$$dU = \left(\frac{\partial U}{\partial S}\right)_V dS + \left(\frac{\partial U}{\partial V}\right)_S dV = TdS - pdV$$

$$dH = TdS + VdP$$

$$dG = -SdT + Vdp$$

$$dA = -SdT - pdV$$

At constant pressure:

$dH = dU + pdV$

At constant temperature:

$dG = dH - TdS$

Derivatives of energy functions:

$$\left(\frac{\partial U}{\partial T}\right)_V = T\left(\frac{\partial S}{\partial T}\right)_V = C_V; \quad \left(\frac{\partial U}{\partial S}\right)_V = \left(\frac{\partial H}{\partial S}\right)_p = T; \quad \left(\frac{\partial U}{\partial V}\right)_S = \left(\frac{\partial A}{\partial V}\right)_T = -p$$

$$\left(\frac{\partial H}{\partial T}\right)_p = T\left(\frac{\partial S}{\partial T}\right)_p = C_p; \quad \left(\frac{\partial H}{\partial p}\right)_S = \left(\frac{\partial G}{\partial p}\right)_T = V; \quad \left(\frac{\partial G}{\partial T}\right)_p = \left(\frac{\partial A}{\partial T}\right)_V = -S$$

Thermodynamic functions for open systems

Internal energy of an open system expressed by internal, extensive variables $U = f(S,V,n_1,n_2, \ldots)$

$$dU = \left(\frac{\partial U}{\partial S}\right)_{V,n_i} dS + \left(\frac{\partial U}{\partial V}\right)_{S,n_i} dV + \left(\frac{\partial U}{\partial n_1}\right)_{S,V,n_{i\neq 1}} dn_1 + \left(\frac{\partial U}{\partial n_2}\right)_{S,V,n_{i\neq 2}} dn_2 + \ldots$$

$$dU = TdS - pdV + \mu_1 dn_1 + \mu_2 dn_2 + \ldots$$

where the chemical potential is defined as $\mu_i = \left(\frac{\partial U}{\partial n_i}\right)_{S,V,n_{j\neq i}}$

Equivalent definitions of chemical potential:

$$\mu_i = \left(\frac{\partial U}{\partial n_i}\right)_{S,V,n_{j\neq i}} = \left(\frac{\partial H}{\partial n_i}\right)_{S,p,n_{j\neq i}} = \left(\frac{\partial G}{\partial n_i}\right)_{p,T,n_{j\neq i}} = \left(\frac{\partial A}{\partial n_i}\right)_{V,T,n_{j\neq i}}$$

The number of components in a mixture is the *minimum* number of pure, neutral compounds needed to compose the mixture. The amounts of the components, n_1, n_2, . . ., can be varied independently.

Partial molar quantities $Y_i = \left(\frac{\partial Y}{\partial n_i}\right)_{p,T,n_{j\neq i}}$ (e.g. $\mu_i = G_i$)

where Y can be any extensive variable in a mixture, e.g. U, H, G, A, S, V.

The total differential of Gibbs energy:

$$dG = dU - TdS - SdT + pdV + Vdp$$

or

$$dG = -SdT + Vdp + \mu_1 dn_1 + \mu_2 dn_2 + \ldots$$

It may be integrated for constant intensive variables, T, p, $x_i = n_i/\Sigma n$:

$$G = \mu_1 n_1 + \mu_2 n_2 + \ldots$$

This generally valid equation can be differentiated for arbitrary changes in composition, and the total differential can be expressed as

$$dG = \mu_1 dn_1 + n_1 d\mu_1 + \mu_2 dn_2 + n_2 d\mu_2 + \ldots$$

Combining the two expressions for the total differential of Gibbs energy, the Gibbs–Duhem equation is obtained:

$$SdT - Vdp + n_1 d\mu_1 + n_2 d\mu_2 + \ldots = 0$$

A similar treatment can be given to other extensive variables.

APPENDIX 2

The Onsager Reciprocal Relations

For a set of homogeneous flux equations:

$$J_i = \Sigma L_{ij} X_j \tag{A2.1}$$

where fluxes are expressed as linear functions of independent forces, the following relations exist between cross coefficients:

$$L_{ij} = L_{ji} \tag{A2.2}$$

These relations were derived by Onsager[1,2] in 1931 and they are called the *Onsager Reciprocal Relations* (*ORR*). Preceding Onsager's general treatment, reciprocal relations had been studied for special cases by Thomson[3] in 1854, Helmholtz[4] in 1876, Nernst[5] in 1888 and Eastman[6,7] in 1926–1928.

Onsager based his treatment on the postulate of *microscopic reversibility*. His treatment is restricted to the *linear range*, where there are linear relations between fluxes and forces. He postulated that the linear relationship is also valid on the microscopic level.

Onsager's theory has been scrutinized and extended by, among others, Casimir,[8] de Groot and Mazur[9] and Meixner.[10,11]

The present simplified derivation of the relations gives the main points without demanding that the reader has an extensive background in mathematics and statistical mechanics. Alternative presentations are given by, for example, Kreuzer[12] and Katchalsky and Curran.[13]

A2.1 *Fluctuations from Equilibrium. Forces and Fluxes*

We shall study fluctuations of thermodynamic functions within an isolated system. The system may be characterized by constant macroscopic values of thermodynamic functions, temperature, pressure, volume and composition. Inside the system there may be momentary local variations, e.g. in composition, and fluctuations around the equilibrium value.

The isolated system will be considered as consisting of two open subsystems. Subsystem 1 is a large reservoir, while subsystem 2 is small and completely surrounded by subsystem 1. The two systems are in equilibrium. The equilibrium is dynamic and fluctuations from equilibrium continue all the time. We shall focus on the fluctuations of thermodynamic variables in the small subsystem 2.

The entropy of a system can be expressed by a set of *independent extensive variables*, A_i:

$$S = f(A_0, A_1, A_2, \ldots) \tag{A2.3}$$

As variables we may choose the internal energy, *U*, the volume, *V*, and the number of moles of components.

For an extensive variable, such as *S*, we have in analogy with the Gibbs–Duhem equation (Appendix 1):

$$A_0 d(\partial S/\partial A_0) + A_1 d(\partial S/\partial A_1) + A_2 d(\partial S/\partial A_2) + \ldots = 0 \tag{A2.4}$$

The *intensive variables*, $\partial S/\partial A_i$, are related by eq. (A2.4). We may choose $\partial S/\partial A_0$ as a function of the other intensive variables, which are then *independent of one another*.

We choose to limit subsystem 2 by a constant value, A_0, for one component, e.g. the solvent, while the other extensive variables may vary independently. Their deviation from the equilibrium value, $A_{i,eq}$ is denoted as α_i and $\alpha_i = A_i - A_{i,eq}$. Since $A_{i,eq}$ is a constant we have $d\alpha_i = dA_i$. As a result of the deviations, the entropy of the isolated system may deviate from the equilibrium value. The deviation, ΔS, can be written as $\Delta S = S - S_{eq}$. Since S_{eq} is a constant, $d\Delta S = dS$. (It can be shown that the deviation in entropy is determined by the second derivatives of *S* with respect to α_i at equilibrium[8,12,13].)

The *forces* in the direction out of subsystem 2 may be defined:

$$X_i = -(\partial \Delta S/\partial \alpha_i)_{\alpha_{j \neq 1}} = -(\partial S/\partial A_i)_{A_{j \neq i}} \tag{A2.5}$$

The conjugate *fluxes* may be defined:

$$J_i = -\,d\alpha_i/dt = -\,dA_i/dt \tag{A2.6}$$

With the limits chosen for subsystem 2 (constant A_0) we have $J_0 = 0$, while J_1, J_2, . . . are independent fluxes. The force X_0, which is a function of the other forces, has no conjugate flux, while X_1, X_2, . . . are independent forces.

A2.2 *Forces and Probability*

The entropy of a system is

$$S = k \ln W \tag{A2.7}$$

where k is the Boltzmann constant and W is the number of *realizations* of the state, i.e. the number of *microscopic states* corresponding to the *macroscopic state*. For the equilibrium state we can write

$$S_{eq} = k \ln W_{eq} \tag{A2.8}$$

The deviation in entropy from the equilibrium value, ΔS, can be written as

$$\Delta S = S - S_{eq} = k \ln (W/W_{eq}) = k \ln P \tag{A2.9}$$

where P is the *probability* of finding the system in a state with entropy S. The antilogarithm of eq. (A2.9) gives

$$P = \mathcal{K} \exp (\Delta S/k) \tag{A2.10}$$

where $\mathcal{K}$ is a *normalization factor* introduced to satisfy the requirement

$$\int \ldots \int P d\alpha_1 d\alpha_2 \ldots = 1 \tag{A2.11}$$

The equilibrium state has the highest probability because this state has the largest number of microscopic states, W_{eq}. Correspondingly the entropy has the maximum value, S_{eq}, at equilibrium. The deviation, ΔS, is therefore *always negative*. Large numerical values of ΔS are very improbable according to eq. (A2.10).

From eqs. (A2.5) and (A2.9) we obtain

$$X_i = -k\frac{1}{P}(\partial P/\partial \alpha_i) \tag{A2.12}$$

A2.3 *Fluxes and Microscopic Reversibility*

Transport in both directions across the border between subsystem 1 and subsystem 2 gives fluctuations in the values of extensive variables. Suppose that α_i has the value α_i' at a given time, t, followed by the event that α_j has the value α_j' after the time Δt. The assumption of *microscopic reversibility* requires that, *over a long time*, this sequence of events will occur equally many times as the sequence $\alpha_j = \alpha_j'$ followed by $\alpha_i = \alpha_i'$ after Δt seconds. The two events can be illustrated on a time-scale:

$\alpha_i = \alpha_i'$ $\qquad\qquad$ $\alpha_j = \alpha_j'$

——— t ——— $t+\Delta t$ ———→

$\alpha_j = \alpha_j'$ $\qquad\qquad$ $\alpha_i = \alpha_i'$

We see that the second event corresponds to the first one if time is reversed.

Mathemetically the postulate of microscopic reversibility may be formulated as

$$\left\langle \alpha_{i,t}\alpha_{j,t+\Delta t} \right\rangle_{av} = \left\langle \alpha_{j,t}\alpha_{i,t+\Delta t} \right\rangle_{av} \tag{A2.13}$$

where $\langle\rangle_{av}$ means time average. We may subtract the same quantity, $\alpha_{i,t}\alpha_{j,t}$, on both sides of eq. (A2.13) and we obtain

$$\left\langle \alpha_{i,t}(\alpha_{j,t+\Delta t}-\alpha_{j,t}) \right\rangle_{av} = \left\langle \alpha_{j,t}(\alpha_{i,t+\Delta t}-\alpha_{i,t}) \right\rangle_{av} \tag{A2.14}$$

We divide both sides of eq. (A2.14) by Δt and we allow Δt to become very small. 'Very small' means much smaller than the relaxation time for fluctuations. On the other hand, Δt should be large compared to the duration of a single collision process. With these reservations we may write

$$\left\langle \alpha_i \mathrm{d}\alpha_j/\mathrm{d}t \right\rangle_{av} = \left\langle \alpha_j \mathrm{d}\alpha_i/\mathrm{d}t \right\rangle_{av} \qquad \text{(A2.15)}$$

For simplicity the subscript t was omitted. The regression of the fluctuations, $\mathrm{d}\alpha_i/\mathrm{d}t$ (or $\mathrm{d}\alpha_j/\mathrm{d}t$), was defined as fluxes (eq. (A2.6)).

Onsager postulated that *the fluxes are linear homogeneous functions of the forces*. Introducing the expression for J_i given in eq. (A2.1) we obtain

$$\left\langle \alpha_i \Sigma_k L_{jk} X_k \right\rangle_{av} = \left\langle \alpha_j \Sigma_k L_{ik} X_k \right\rangle_{av} \qquad \text{(A2.16)}$$

A2.4 *The Reciprocal Relations between Cross Coefficients*

In order to find the consequences of eq. (A2.16), we shall compare the time average of the factor $\alpha_i X_k$ on the left-hand side with the time average of the factor $\alpha_j X_k$ on the right-hand side.

A time average can be expressed by means of probability. The probability of a particular state of a system with variables within narrow limits, α_1 in the interval α_1 to $\alpha_1 + \mathrm{d}\alpha_1$, α_2 in the interval α_2 to $\alpha_2 + \mathrm{d}\alpha_2$, etc., is equal to the product $P\mathrm{d}\alpha_1\mathrm{d}\alpha_2 \ldots$. The probability of the system being in any one of the possible states is equal to unity and it can be expressed by eq. (A2.11). Over (a long) time a system will go through all possible states. It is a postulate that the fraction of time that the system is in the particular state can be expressed by the same product $P\mathrm{d}\alpha_1\mathrm{d}\alpha_2 \ldots$. The sum of the time fractions spent in all possible states is equal to unity as expressed by eq. (A2.11). The time average of $\alpha_i X_k$ is obtained by multiplying $\alpha_i X_k$ in a given state by the fraction of time spent in this state and then summarizing these products over a long time. Thus we obtain

$$\left\langle \alpha_i X_k \right\rangle_{av} = \int \ldots \int (\alpha_i X_k) P \mathrm{d}\alpha_1 \mathrm{d}\alpha_2 \ldots \qquad \text{(A2.17)}$$

X_k may be replaced by its expression according to eq. (A2.12):

$$\left\langle \alpha_i X_k \right\rangle_{av} = -k \int \ldots \int \alpha_i(\partial P/\partial \alpha_k) d\alpha_1 d\alpha_2 \ldots d\alpha_k \ldots \quad \text{(A2.18)}$$

The term $\alpha_i(\partial P/\partial \alpha_k)d\alpha_k$ may be subjected to integration by parts:

$$\int_{-\infty}^{+\infty} \alpha_i(\partial P/\partial \alpha_k) d\alpha_k = \Big|_{-\infty}^{+\infty} \alpha_i P - \int_{-\infty}^{+\infty} P(\partial \alpha_i/\partial \alpha_k) d\alpha_k \qquad \text{(A2.19)}$$

The first term on the righthand side of the equation is equal to zero, because the probability of finding infinite (or very large) fluctuations is equal to zero. Further, since α_i and α_k represent independent deviations from equilibrium values, $\partial\alpha_i/\partial\alpha_k = 0$ when $k \neq i$ and $\partial\alpha_i/\partial\alpha_k = 1$ when $k = i$. Hence

$$\left\langle \alpha_i X_k \right\rangle_{av} = 0, \text{ for } k \neq i \qquad \text{(A2.20)}$$

$$\left\langle \alpha_i X_k \right\rangle_{av} = k, \text{ for } k = i \qquad \text{(A2.21)}$$

This means that all terms on the left-hand side of eq. (A2.16) are zero, except the term where $k=i$. On the right-hand side all terms, except the one where $k=j$, are zero. We are left with the terms

$$kL_{ji} = kL_{ij} \qquad \text{(A2.22)}$$

or

$$L_{ji} = L_{ij} \qquad \text{(A2.23)}$$

which represents the *Onsager Reciprocal Relations*.

An important basis for the above development is independent fluxes. There is one more force than there are fluxes, the dependent force $-\partial S/\partial A_0$. Since this force has no conjugate flux, the coefficients L_{0i} are equal to zero. According to the Onsager reciprocal relations, the coefficients L_{i0} are also equal to zero. Thus this force does not affect the fluxes and only independent forces enter the flux equations.

It has been shown by de Groot and Mazur[9] that if forces are independent, but fluxes interdependent, eq. (A2.23) is still valid. With both interdependent fluxes and interdependent forces the Onsager relations are not necessarily fulfilled. It is possible, however, to choose the coefficients in such a way that the Onsager relations hold.

The Onsager relations were developed for the fluctuations in a system at equilibrium. They are assumed to be also valid for a system not at equilibrium, if we have *local equilibrium*. Local equilibrium implies that the rate of a directed movement is small compared to the fluctuations (see Section 1.1).

When magnetic forces or Coriolis forces are involved the reciprocal relations are $L_{ij} = -L_{ji}$.[1,2]

A2.5 *Thomson's Hypothesis and the Onsager Reciprocal Relations*

When Thomson[3] derived the relation between the Peltier heat and the thermoelectric power (eq. (6.23)) he introduced the hypothesis that the process could be considered as consisting of two parts, an irreversible process and a reversible one. The latter process could be treated by equilibrium thermodynamics. We shall illustrate the connection between Thomson's hypothesis and the Onsager reciprocal relations by means of an example, the transport of charge and mass in a simple electrochemical cell (cf. Section 4.1.1):

$$\mathrm{Ag(s)|AgCl(s)|HCl(aq,}c_\mathrm{I}\mathrm{)||HCl(aq,}c_\mathrm{II}\mathrm{)|AgCl(s)|Ag(s)} \qquad \text{(A2.24)}$$

The dissipation function in the liquid junction is,

$$T\Theta = -J_1\nabla\mu_1 - j\nabla\varphi \qquad \text{(A2.25)}$$

where J_1 is the flux of HCl with H_2O as frame of reference and $\nabla\mu_1$ is the gradient in chemical potential for HCl. The flux equations are (cf. eq. (4.4))

$$J_1 = -L_{11}\nabla\mu_1 - L_{12}\nabla\varphi \qquad \text{(A2.26a)}$$

$$j = -L_{21}\nabla\mu_1 - L_{22}\nabla\varphi \qquad \text{(A2.26b)}$$

By rearrangement of eq. (A2.26b) we have

$$\nabla\varphi = -(L_{21}/L_{22})\nabla\mu_1 - (1/L_{22})j \qquad (A2.27)$$

Eliminating $\nabla\varphi$ from eqs. (A2.26) we obtain (cf. eq. (4.9))

$$J_1 = -(L_{11} - L_{12}L_{21}/L_{22})\nabla\mu_1 + (L_{12}/L_{22})j \qquad (A2.28)$$

Replacing J_1 and $\nabla\varphi$ in eq. (A2.25) by the above expressions, we obtain (cf. Meixner[11])

$$T\Theta = \underset{(a)}{(L_{11} - L_{12}L_{21}/L_{22})(\nabla\mu_1)^2} + \underset{(b)}{(1/L_{22})j^2}$$
$$\underset{(c)}{-(L_{12}/L_{22})j\nabla\mu_1} + \underset{(d)}{(L_{21}L_{22})\nabla\mu_1 j} \qquad (A2.29)$$

The expression for $T\Theta$ above contains four terms marked (a), (b), (c) and (d) successively.

The phenomenological coefficients are independent of the forces and the forces are independent of each other. We shall choose a situation where $\nabla\mu_1$ is time independent, i.e. we have the stationary state for diffusion. We shall also have $j=0$. When the emf of a cell is measured, we have approximately these conditions. The current density varies around zero, and large reservoirs on both sides of the liquid junction permit an approximately stationary state for diffusion.

We shall identify the four terms in eq. (A2.29). Their dependence on j is shown in fig. A2.1.

The term (a) represents the constant contribution to $T\Theta$ from diffusion of HCl. The term (b) represents the contribution from Joule heat. It is proportional to j^2 and decreases rapidly with decreasing numerical values of j. The terms (a) and (b) are always positive. The term (c) represents the change in Gibbs energy due to transport of electric charge. The term (d) represents the electric work supplied less the contribution from Joule heat. Terms (c) and (d) are both reversible. They change sign when j changes sign.

Introducing the Onsager reciprocal relation $L_{12} = L_{21}$ we see that the two terms (c) and (d) cancel:

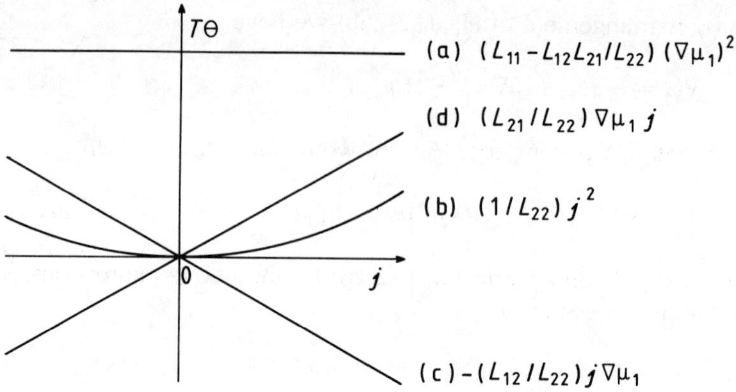

Figure A2.1. The contributions to the dissipation function for current densities near zero.

$$\overset{(c)}{-(L_{12}/L_{22})j\nabla\mu_1} + \overset{(d)}{(L_{21}L_{22})\nabla\mu_1 j} = 0 \tag{A2.30}$$

The sum of the two reversible terms does not contribute to $T\Theta$. This means that the reversible part of the total process can formally be treated as an equilibrium process.

The reversible part of the process can be treated in the following way. For $j=0$, eq. (A2.27) reduces to $\nabla\varphi = -(L_{21}/L_{22})\nabla\mu_1$. By the Onsager reciprocal relation we have $L_{21}/L_{22} = L_{12}/L_{22}$. The last fraction is equal to the transference coefficient for component 1 (HCl), and we have

$$\nabla\varphi = -t_1\nabla\mu_1 \tag{A2.31}$$

This equation may be integrated over the cell:

$$\Delta\varphi = -\int_{(I)}^{(II)} t_1 d\mu_1 = \Big|_{(I)}^{(II)} t_1\mu_1 + \int_{(I)}^{(II)} \mu_1 dt_1 = -\Delta G(Q) \tag{A2.32}$$

where $\Delta G(Q)$ is the change in Gibbs energy of the cell due only to the *reversible charge transfer* (cf. the derivation in Section 4.2.2 leading to eq. (4.82)).

By means of an example the Onsager reciprocal relations have been used to confirm Thomson's hypothesis. When solving problems in irreversible thermodynamics it is often convenient to consider a process as consisting of a reversible process and an irreversible one.

A2.6 *References*

1. Onsager, L., *Phys. Rev.*, **37**, 405 (1931).
2. Onsager, L., *Phys. Rev.*, **38**, 2265 (1931).
3. Thomson, W. (Lord Kelvin), *Proc. Roy. Soc. Edinburgh*, **3**, 225 (1854).
4. v. Helmholtz, H., *Wied Ann.*, **3**, 201 (1876).
5. Nernst, W., *Zeits. f. physik. Chem.*, **2**, 613 (1888).
6. Eastman, E.D., *J. Am. Chem. Soc.*, **48**, 1482 (1926).
7. Eastman, E.D., *J. Am. Chem. Soc.*, **50**, 283, 292 (1928).
8. Casimir, H.B.G., *Rev. Mod. Phys.*, **17**, 343 (1945).
9. de Groot, S.R. and Mazur, P., *Non-Equilibrium Thermodynamics*, North-Holland, Amsterdam, 1962.
10. Meixner, J., *Ann. Phys.*, **43**, 244 (1943).
11. Meixner, J., *Adv. Mol. Relax. Processes*, **5**, 319 (1973).
12. Kreuzer, H.J., *Nonequilibrium Thermodynamics and its Statistical Foundation*, Clarendon Press, Oxford, 1981.
13. Katchalsky, A. and Curran, P.F., *Nonequilibrium Thermodynamics in Biophysics*, Harvard University Press, 1981.

APPENDIX 3

Exercises—Answers to Numerical Problems

1.1. (i) -5×10^{-5}%; (ii) -5×10^{-3}%; (iii) -0.5%; (iv) -9%.
1.2. (a) (v) 1; (b) (v) 2; (c) (v) 1.
1.3. (c) $c_{H^+,\ell} = 0.07$, $c_{K^+,\ell} = 0.13$, $c_{H^+,r} = 0.03$, $c_{K^+,r} = 0.07$, all kmol m^{-3}.
3.2. (a) 2; (b) 1; (c) 2.
4.3. (a) 0.059 V; (b) -0.020 V.
4.4. (c) 0.385.
4.5. (b) 2.
5.1. 1.4.
5.2. (a) 0.118 V; (b) 0 V.
5.3. (i) $\Delta\varphi_1 = 0$, $\Delta\varphi_2 = 0$; (ii) $\Delta\varphi_1 = -18$, $\Delta\varphi_2 = -399$; (iii) $\Delta\varphi_1 = -40$, $\Delta\varphi_2 = -570$. Dimension: J faraday^{-1}
5.4 -11.0 J faraday^{-1}
5.5. $t_{H_2O} = 1.6$.
5.7. (d) $l_{22} = 0.55\times10^{-12}$ $mol^2\ J^{-1}s^{-1}m^{-1}$.
6.1. (a) $s_T = 2.5\times10^{-3}\ K^{-1}$, $q^* = 1.8$ kJ mol^{-1}; (b) $J_{KCl} = -9.5\times10^{-8}$ mol $s^{-1}\ m^{-2}$.
6.2. $\tau_{Pt} = -\ 0.01146\ T - 229\ T^{-1} + 17544\ T^{-2}$ (J K^{-1} faraday^{-1}).
6.3. (b) $t_{K^+}S^*_{K^+} - t_{Cl^-}S^*_{Cl^-} = 7.0$ J K^{-1} faraday^{-1}; (c) $\Delta\varphi/\Delta T = 61.0$ J K^{-1} faraday^{-1}, $S^*_{K^+} = 128$ J K^{-1} faraday^{-1}, $S^*_{Cl^-} = 114$ J K^{-1} faraday^{-1}.
8.1. (c) $J_{H^+} = 3.9\times10^{-7}$ mol $m^{-2}\ s^{-1}$.
8.2. (a) 27.0 mV; (b) 27.3 mV.
10.1. (a) 0.81 kT; (b) $v/v_{linear} = 0.83/0.81$.
10.2. (a) (i) -167 mV, (ii) -222 mV, (iii) -334 mV; (b) (i) -347 mV, (ii) -231 mV, (iii) -174 mV; (c) $m = 3$ and $n = 6$, $m = 4$ and $n = 6$ or 9.

APPENDIX 4

List of Symbols

Common thermodynamic symbols, used and defined in Appendix 1, are not listed here. Numbers in parentheses refer to the page where the symbol is first introduced and to the page where it is defined.

$\lvert^{A}\rvert$	Anion exchange membrane (106, 182)
A	Contribution to emf (163)
$\mathrm{A_i}$	Extensive variable (261)
A	Cross-section (62)
A	Affinity of chemical reaction (217)
a	Acceleration (136)
a	Activity (51)
B	Constant in extended Debye–Hückel equation (171)
$\lvert^{C}\rvert$	Cation exchange membrane (106, 182)
c	Concentration (41)
D	Fick's diffusion coefficient (52)
D_T	Thermal diffusion coefficient (111)
D^S	Fick's diffusion coefficient, solvent-fixed reference (138)
D^V	Fick's diffusion coefficient, volume-fixed reference (139)
E	Electromotive force, emf (unit, *V*) (43)
E_a	Energy barrier, activation energy (5)
e	Electric charge of an electron (5)
F	Number of degrees of freedom (96)
F	Faraday constant (43)
f	Activity coefficient, mole fraction basis (72)
d*G*(*Q*)	Charge-dependent change in Gibbs energy (63)

$dG(t)$	Time-dependent change in Gibbs energy (63)
h	Level, altitude (135)
I	Electric current, flux of charge (37)
I	Ionic strength (171)
J	Flux (18)
J_D	Diffusion flux (88)
J_i	Flux of component i (25)
J_q	Measurable heat flux (25)
J_S	Flux of entropy (26)
J_V	Volume flux (81)
J_Φ	Total heat flux (108)
J^B	Flux, barycentre-fixed frame of reference (103)
J^M	Flux, membrane-fixed frame of reference (104)
J^S	Flux, solvent-fixed frame of reference (137)
J^V	Flux, volume-fixed frame of reference (139)
j	Current density, flux of charge through unit area (7)
K	Equilibrium constant (72)
K_w	Equilibrium constant, water dissociation product (186)
K	Constant in Harned's rule (151)
$\mathcal{K}$	Normalization factor (262)
k	Boltzmann constant (5)
k	Parameter (201)
k_1	Rate constant (5)
L, L_{ij}	Phenomenological coefficient (29)
L_{ii}	Direct coefficient (30)
L_{ij}	Cross coefficient, $j \neq i$ (30)
$\mathcal{L}$	Composite coefficient (81)
l, l_{ij}	Diffusion coefficient (35, 46)
ℓ	Length (of cell, system, interval etc.) (38)
ℓ	Left-hand side (13,187)
$\|\|$	Liquid junction (41)
M^-	Cation site in membrane (71, 73)
M_i	Molar mass of component i (103)
m_i	Mass of component i (135)
n	Number of components (25)
n_i	Amount of component i (number of moles) (15)
P	Number of phases (96)
P	Probability (262)
Q	Quantity of electricity, electric charge (22)
q_i^*	Heat of transfer, component i (112)

R	Gas constant (11)
R	Electric resistance (64)
R	Resistance coefficient (220)
r	Right-hand side (13)
r	Radius (140)
S^*	Entropy transferred per faraday (117)
S^*_{A}	Entropy transported through conductor A (119)
$S^*_{\mathrm{X}^-}$	Entropy transported by ion X^- (124)
$\hat{S}_{A^+}$	Eastman entropy of ion A^+ (132)
$\mathcal{S}^{S}$	Sedimentation coefficient, solvent-fixed frame of reference (138)
$\mathcal{S}^{V}$	Sedimentation coefficient, volume-fixed frame of reference (139)
s_T	Soret coefficient (112)
Ten	Tension (muscle) (217)
$T\Theta$	Dissipation function (25)
t	Time (17, 31)
t_{i}	Transference number of the ion i or transference coefficient of the neutral component i (42, 44)
u_{i}	Electric mobility of the ion i (50)
v	Rate of transport, velocity (5)
v_{i}	Specific volume of component i (136)
W	Number of realizations of a state (262)
X	Force (18)
X_{i}	Driving force for flux of component i (26)
X_j	Driving force for electric current density (26)
X_q	Driving force for heat flux (26)
X_S	Driving force for entropy flux (26)
x	Distance in the x-direction (6)
x_{i}	Mole fraction of component i or equivalent fraction of ion i (33, 199)
y	Activity coefficient, concentration basis (51)
z_{i}	Charge of the ion i (53)
α	Coupling coefficient (35)
α	Deviation from equilibrium value (261)
β	Asymmetry factor (7)
η	Overpotential (7)
η	Efficiency (17)
η	Viscosity (140)
Θ	Entropy production per unit volume and unit time (25)

κ	Electric conductivity (44)
λ	Thermal conductivity (111)
μ_i	Chemical potential of component (species, ion) i (7, 258)
$\tilde{\mu}_i$	Electrochemical potential of single ion (53)
$\nabla\mu_i(c)$	Gradient in chemical potential caused by gradient in concentration only (79)
$\Delta\mu_{i,T}$	Change in chemical potential at constant temperature (21)
ν_i	Stoichiometric coefficient for species i (8)
Π	Osmotic pressure (84)
π	Peltier heat (116)
ρ	Density (103)
σ	Reflection coefficient (89)
τ	Thomson coefficient (119)
$d\Phi$	Total heat transferred (19)
$\Delta\varphi$	Electric potential (unit, J faraday^{-1}) (5, 23)
$\nabla\varphi$	Gradient in electric potential (26)
$\Delta\varphi_{obs}$	Observed electric potential (79)
ψ	Electrostatic potential (53)
ω	Angular velocity (136)

INDEX

F

G